T0136846

Photon Counting Detectors for X-ray Imaging

Hiroaki Hayashi • Natsumi Kimoto
Takashi Asahara • Takumi Asakawa
Cheonghae Lee • Akitoshi Katsumata

Photon Counting Detectors for X-ray Imaging

Physics and Applications

 Springer

Hiroaki Hayashi
Kanazawa University
Kanazawa, Ishikawa, Japan

Natsumi Kimoto
Kanazawa University
Kanazawa, Ishikawa, Japan

Takashi Asahara
Kanazawa University
Kanazawa, Ishikawa, Japan

Takumi Asakawa
Kanazawa University
Kanazawa, Ishikawa, Japan

Cheonghae Lee
Kanazawa University
Kanazawa, Ishikawa, Japan

Akitoshi Katsumata
Asahi University
Gifu, Japan

ISBN 978-3-030-62682-2   ISBN 978-3-030-62680-8   (eBook)
https://doi.org/10.1007/978-3-030-62680-8

This Springer imprint is published by the registered company Springer Nature Switzerland AG
The registered company address is: Gewerbestrasse 11, 6330 Cham, Switzerland

Introduction

Currently, energy-resolving photon counting detectors are the focus of attention as the next-generation-type imaging detector. One application is the production of quantitative medical images, which can lead to novel diagnosis through medical X-ray examination. To achieve this purpose, we should analyze spectra measured with an energy-resolving photon counting detector. However, actual measured spectra are not ideal because they are affected by the physics of the detector. In this book, we will describe the importance of taking into consideration the physics when developing a novel imaging detector and propose analytical procedures to correct the problems that occur. Also, we will present the first results of a novel effective atomic number image produced with a prototype photon counting imaging detector.

First, we will explain the impact of the development of a novel photon counting detector on the medical field. The key point of this book is understanding an "energy-resolving photon counting detector (ERPCD)", which is a detector which can analyze each X-ray independently. In contrast, an "energy integrating detector (EID)" is conventionally used for current clinical X-ray diagnosis. Second, in order to understand the difference between the ideal X-ray spectrum and actual measured spectrum, it is important to analyze the physical phenomena which occur in the imaging detector. To help with this explanation, we will explain basic physics and the explanation will also extend to the analysis of the physics of a multi-pixel-type photon counting detector. Next, we will demonstrate our novel analytical procedure to exemplify the potential of deriving a quantitative image; at this time, we will present effective atomic number images that we have produced. When we understand both X-ray attenuations in the object and physics of the detector, we can precisely analyze the X-ray spectra which are measured with a photon counting detector. Finally, we will show you examples of effective atomic number images of ordinary objects.

Contents

1 Generation of X-rays 1
1.1 X-rays Using for Imaging 1
1.2 What Is an X-ray? 1
1.3 Atomic Model 4
1.4 Multielectron Atom 8
1.5 Historical Experiment to Verify the Particle Nature of Photons ... 11
1.6 Generation of X-rays 13
    1.6.1 Interactions between Energetic Electrons
        and Target Material 13
    1.6.2 Characteristic X-rays 17
    1.6.3 Polychromatic X-rays 18
    1.6.4 Attenuation of X-rays 21
1.7 X-ray Spectrum 23
1.8 Application of X-rays 27
References 29

2 Introduction to Physics in Medical X-ray Diagnosis 31
2.1 X-ray Imaging 31
2.2 Medical X-ray Diagnosis 34
2.3 Energy Integrating Detector (EID) and Energy Resolving
    Photon Counting Detector (ERPCD) 36
2.4 What Can We Analyze from the Images Measured
    with ERPCD? 37
2.5 Review of Current Novel Analysis Procedure Concerning
    the Analysis of Objects 40
2.6 Outline to Derive Precise Material Identification
    Using an ERPCD 42
References 43

3 Radiation Detector Physics 45
   3.1 What Is Response Function? 45
   3.2 Interaction Provability of Semiconductor Materials Used
       in the ERPCD. 50
   3.3 Interactions Between Incident X-rays and Atoms 52
       3.3.1 Coherent Scattering 52
       3.3.2 Photoelectric Effect 53
       3.3.3 Compton Scattering 55
   3.4 Response Functions and X-ray Spectra 59
       3.4.1 Response Function of a Single-Probe-Type
             CdTe Detector 59
       3.4.2 Response Function of a Multi-Pixel-Type
             CZT Detector. 64
       3.4.3 Consideration of Charge Sharing Effect
             and Energy Resolution. 72
       3.4.4 Reproduction of X-ray Spectra Measured with
             a Multi-Pixel-Type CZT Detector 74
   3.5 Polychromatic X-rays and Effective Energy 84
   3.6 Novel Correction Method for Detector Response and Beam
       Hardening Effect 86
   References. 91

**4 Material Identification Method with the Aim
of Medical Imaging**. 93
   4.1 Information Which Can Be Derived from Attenuation Factor. 93
   4.2 Method to Derive Z_{eff} Value Using ERPCD. 100
   4.3 Experimental Verification of Z_{eff} image 109
   References. 114

5 Summary 115
   Reference 116

Index. 117

About the Authors

Hiroaki Hayashi worked for 2 years as an assistant professor at the Radioisotope Research Center, Nagoya University (2009–2011). He is specialized in nuclear physics, and his research theme for his doctorate was atomic mass measurement by detecting radiations emitted from radioisotopes. After receiving his Ph.D., he moved to Tokushima University as an assistant professor and changed his research area to radiation physics in medicine (2011–2017). He then moved to Kanazawa University as an associate professor and continued to carry out work in the research area (2017–present). His current speciality is radiation physics concerning dosimetry and development of next-generation-type imaging detectors. The physics-based analytical methods described in this book largely depend on his ideas.

Natsumi Kimoto has been performing research with the main theme "development of novel plain X-ray imaging system," since she was an undergraduate student at Tokushima University. After receiving her Bachelor's (2016) and Master's (2018) degrees at Tokushima University, she moved to a doctorate course at Kanazawa University (2018–present). To extend the use of our method to clinical environments, she has been analyzing several issues using an actual proto-type detector. The knowledge which was acquired through experimentation and simulation studies was presented at various authoritative international conferences such as IEEE and RSNA. At RSNA 2019, her research was awarded a "Certificate of Merit."

Takashi Asahara earned his Bachelor's degree in 2017 from Tokushima University. He has been participating in Dr. Hayashi's Lab since then. After finishing the undergraduate school he entered a Master's course at Tokushima University. He also started working as a radiological technologist at Okayama University Hospital (2018–present). After earning a Master's degree in 2019, he entered a doctoral course at Kanazawa University (2019–present). His current research theme is concerning the development of a novel dosimeter. He is good at performing Monte-Carlo calculations, and response functions used in this book are based on his research.

Takumi Asakawa enrolled in an undergraduate course at Kanazawa University (2015–2019). He received radiological technologist license in 2019. His Bachelor's research theme was to simulate a photon counting detector's response to various semiconductor materials. After receiving a Bachelor's degree (2019), he entered a Master's course at Kanazawa University (2019–present). The analysis of linear attenuation coefficient described in this book was largely due to his contribution. His current research theme is to optimize detector settings toward medical diagnostic applications.

Cheonghae Lee is a 4th-year undergraduate student at Kanazawa University. This year, she belongs to Dr. Hayashi's Laboratory (2020–present). Thanks to her great contribution, we were able to complete the writing of this book by creating many easy-to-understand drawings. Her research theme is concerning the improvement of the medical image which is obtained with a photon counting detector.

Akitoshi Katsumata obtained his Ph.D. degree in 1993 from Asahi University. Since then, he has specialized in the fields of radiology and dentistry. He became a lecturer in 1995, an associate professor in 1998, and a professor in 2011. He has a dentist license and has a great deal of clinical experience. His research has been the development of photon counting technology based on dental diagnostic devices, including panoramic and intraoral radiography.

Chapter 1
Generation of X-rays

1.1 X-rays Using for Imaging

Examinations using X-rays are widely applied to various industrial and clinical applications. Especially, the use of X-rays that are used to diagnose various diseases has been focused on since the early stages of X-ray development. Because X-ray examinations are non-invasive and have a record of achieving high performance, many current clinical diagnoses are based on X-ray images [1–3]. We understand that many imaging techniques help medical diagnosis and contribute to the improvement of quality of life. Although the X-ray examination is an essential element for medical diagnosis, the principles of generating X-ray images have not made much progress. In this book, we want to describe the development of a novel imaging detector, which is expected to open new avenues for diagnosis in medicine. Before describing the novel analysis technique using the new imaging detector, we will explain the basics of X-rays.

1.2 What Is an X-ray?

As generally known in various textbooks of physics, one of the important characteristics of radiation is it has the ability to cause "ionization." Using our experimental apparatus, we will explain a simple experiment to help you understand the ability to cause ionization. Figure 1.1 shows a special leaf electroscope [4] which was made for an experiment using medical diagnostic X-ray equipment. Figure 1.1a shows the experimental arrangement. The lower part of the leaf electroscope is irradiated with X-rays in a closed position. Figure 1.1b shows a schematic drawing of the leaf electroscope; there are two chambers. In the upper space, there are thin aluminum foils. When the foil is charged, a repulsive force is generated and the foils are opened. In

© The Author(s), under exclusive license to Springer Nature Switzerland AG 2021
H. Hayashi et al., *Photon Counting Detectors for X-ray Imaging*,
https://doi.org/10.1007/978-3-030-62680-8_1

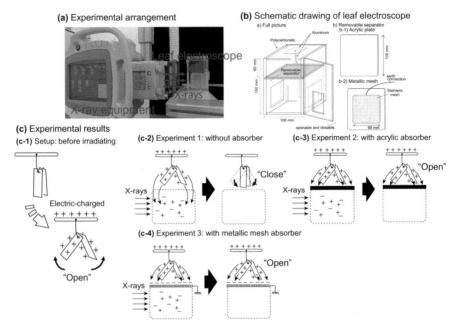

Fig. 1.1 The concept of leaf electroscope for the experiment of leaf electrometer to verify the generation of ions caused by X-ray irradiation

other words, by observing the condition of the foils, we can qualitatively know the charge state of the foils. On the other hand, the lower space is filled with the air, and the X-rays can be introduced from the side windows. One of the unique things of this leaf electroscope is that we can insert a separator to divide the inner space into upper and lower chambers. At this time, we can choose two separators. One is an acrylic plate, which has the effect of physically blocking the movement of gas molecules and electrons, and blocking the electric field from the charges on the foils. The other is a plate having a window with metallic mesh. The mesh can shield only the effect of the electric field without inhibiting the movement of gas molecules and electrons. The experimental results are shown in Fig. 1.1c. (c-1) is the predetermined condition; we put electric charges on the aluminum foils to open the leaves. (c-2) is the result of the condition, in which X-rays are introduced to the lower space and no separators are used; the foil closes as soon as X-rays are exposed. From the schematic drawing in (c-2), we can understand the phenomenon when X-rays are introduced to the air region. Namely, X-ray can ionize the air, and produced electrons which move to the foils to cancel out the charges. (c-3) shows the result of the condition in which an acrylic separator was inserted before X-ray irradiation. We can see that there is no change in the leaves. From this fact, we can understand that the cause of the leaf closing is a phenomenon happening in the lower space. When observing this experiment if we had knowledge concerning the phenomenon of ion-ization caused by X-ray irradiation, we can easily understand the result. However, we want to explain this experiment to someone without previous knowledge of ion-ization, and want to verify the fact that the electrons are generated in the irradiated area and they are transported by the electric field. A second experiment is important to understand this phenomenon. (c-4) shows the results in which a metallic mesh separator was inserted to separate the space. When X-rays are irradiated, the leaves are still opened. Can you explain the meaning of this result? As mentioned above, gas molecules and electrons can move freely through the mesh separator. However, when cutting the electric field with the mesh, there is no longer the situation in which electrons are collected by the electric field. That is why the leaves remain open under these conditions. From these series of experiments, it is possible to infer the following phenomenon; gas molecules are ionized and electrons are created in the area where X-rays are irradiated.

An X-ray is a type of electromagnetic wave. In this section, we will explain the concept of X-ray based on historical verification processes. The explanation of X-ray emission is strongly related to the atomic model. Figure 1.2 is a schematic drawing of an experiment to prove the wave character of electromagnetic waves. The equipment consists of two parts; one is a prism which can separate electromag-netic waves into each wavelength, and the other is a detector. It is easy to imagine color film to use as a detector, but any detector that can measure the number of electromagnetic waves can be substituted. Figure 1.2a shows an experimental result when sunlight is analyzed with this system; since sunlight is a collection of visible lights that vary in energy and continuously change, they are divided into seven col-ors by a prism. The generating mechanism can be explained by the theory of black body radiation. A heated black box emits continuous visible light. In other words, the intensity distribution of the visible light is a function of temperature as shown in the upper right graph in Fig. 1.2a. Interestingly, when deriving this continuous light distribution, it was necessary to think of light as quantum rather than a wave. This theory is one of the most interesting parts of quantum mechanics, and furthermore, research on the light emitted from an excited atom has also provided important experimental data that gives us advance quantum mechanics. Figure 1.2b shows an experimental result when hydrogen gas was used as a light source. It was surprising because only a few lines of the light spectrum can be seen. The visible light of the hydrogen atom observed in this experiment are at wavelengths of 656 nm, 486 nm, 434 nm, and 410 nm. We were able to solve this mysterious sequence of numbers along with the elucidation of the light-generating process by considering the atomic model having a nucleus.

The name X-ray comes from the mechanism of generation. Figure 1.3 shows the relationship between energy (wavelength) and the names of electromagnetic waves. The energy of visible light is from 1.5 to 3 eV. It is amazing that such a narrow electromagnetic wave band provides so many colors and creates a colorful world. Visual information based on visible lights is very useful for recognizing many things. In a similar way, we can expect that careful energy analysis of X-rays will

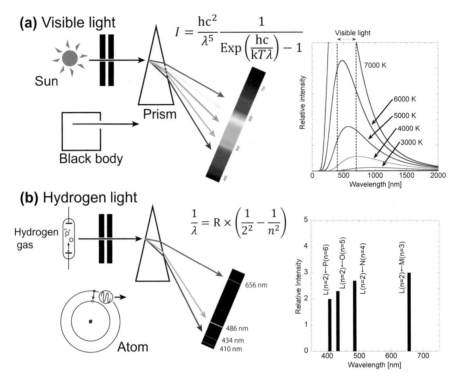

Fig. 1.2 Spectrum comparison of (**a**) visible light and (**b**) hydrogen light. The continuous spectrum can be seen for visible light, but the spectrum having several lines is observed when analyzing the hydrogen light

bring a lot of information. Compared to visible light, γ-ray and X-rays have much higher energies. In general, γ-rays have the characteristic that they have higher energy than X-rays, but the difference in names comes from the way they are generated. Namely, γ-rays are electromagnetic waves emitted from atomic nuclei, and X-rays are emitted from atoms. It should be noted that the X-rays mentioned here are characteristic X-rays, and the X-rays used mainly in medical imaging are bremsstrahlung X-rays which will be explained later. It is an important point that the excited energies of atomic nuclei and atom are unharnessed or released by emitting the electromagnetic waves.

1.3   Atomic Model

Understanding atomic structure is essential for understanding X-rays. This section describes Bohr's atomic model. Although this model can explain the concept of a bound electron in atoms, it should be noted that this simple calculation can only be

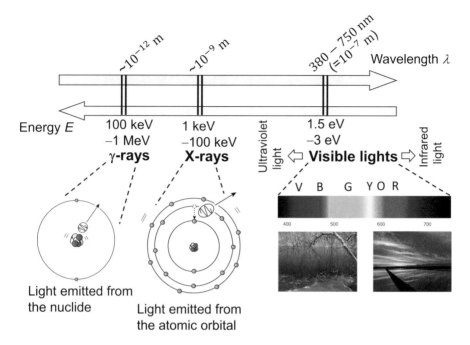

Fig. 1.3 The relationship between energy and wavelength of an electromagnetic wave. Visible light ranges between 1.5 and 3.0 eV. The X-rays and γ-rays are in the 1–100 keV and 100 keV–1 MeV, respectively. It should be noted that the difference between X-rays and γ-rays depends on where they are emitted. The amount of energy is only a rough guide

applied to a two-body model consisting of one electron and the nucleus. When more complicated calculations such as the orbits of the atoms having many bounded electrons are required, it is necessary to solve the Schrödinger equation. Bohr's atomic model played a very important role in the introduction of quantum mechanics in the twentieth century. He succeeded to calculate the radius and energy state of electrons by introducing a new concept to the classical theory of quantum mechanics.

Figure 1.4 shows two conditions which were additionally introduced to Bohr's theory. The first condition is called the "quantum condition"; the electrons can exist in the orbitals under limited situations. Namely, as shown in Fig. 1.4a, it was considered that only a standing wave can exist in the orbital when an electron was considered to be a particle with wave nature. Under these conditions, only discrete orbits that are an integral multiple of the standing wave are allowed. The number of standing waves determined at this time, n, is called the main quantum number and is a very important. The second condition is called the "frequency condition." Under this condition, the basic concept was shown; characteristic X-rays can be emitted when an electron is de-excited from a high energy state orbital to a low energy state orbital. When applying this theory to the hydrogen atom, as shown in Fig. 1.4b, a Lyman series to $n = 1$ orbital transition, a Balmer series to $n = 2$ transition and a Paschen series to $n = 3$ transition were observed. Among these transitions, it was also found that the visible light shown in Fig. 1.3 is in the Balmer series.

Bohr's atomic model for Z=1 atom

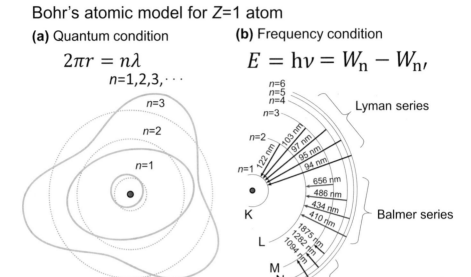

(a) Quantum condition

$$2\pi r = n\lambda$$
$$n=1,2,3,\cdots$$

(b) Frequency condition

$$E = h\nu = W_{n} - W_{n'}$$

Fig. 1.4 Bohr's model of Z = 1 atom, which has only one electron in its orbital. In order to calculate the radius and energy state of the electron, we should consider the following two conditions: (**a**) quantum condition and (**b**) frequency condition

Based on Bohr's model, the radius of the bounded electron and the binding energy can be calculated as shown in Fig. 1.5. It is well known that the radius of electron shells increases rapidly as the quantum number increases as shown in Fig. 1.5a. The electron orbits have names, and the names of orbitals corresponding to n = 1,2, and 3 are called K shell, L shell, and M shell, respectively. Among these orbitals, the K shell is especially important, because most of the interactions between X-rays and matter that we deal with in this book are due to K-shell electrons. The chemical properties of the elements are determined by the properties of the outermost shell electrons, but for radiation physics, the behavior of K-shell electrons is the most important. Figure 1.5b shows binding energies of electrons bounded in the orbitals; the binding energy is defined by the sum of Coulomb potential and the kinetic energy of the electron. As clearly seen in the graph, binding energies of electrons are negative values. The state where n = ∞ means that the electron is unbound by the Coulomb potential of the nucleus and ionized. In this case, the binding energy becomes zero. In other words, the situation where the electron is bound in the electron orbital can be expressed by the mathematical description which shows a negative binding energy.

Another major development in the twentieth century was the theory of relativity which was discovered by Einstein. Studies of this theory have shown that mass is equivalent to energy. The discovery developed the concept of classic physics, which was used to analyze kinematics by describing forces, and established the notion of understanding physical phenomena by considering energy states. Here we will show you an example of ionization. Figure 1.6 shows a comparison of the energy

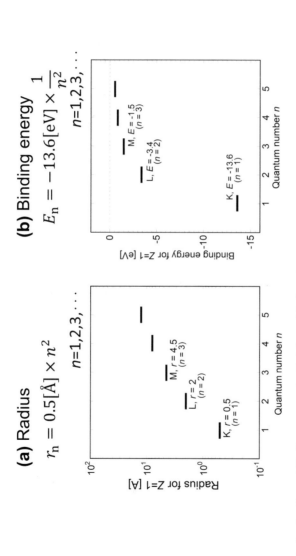

Fig. 1.5 The calculated radius and binding energy using Bohr's model of $Z = 1$ atom. These values can be expressed as a function of quantum number n

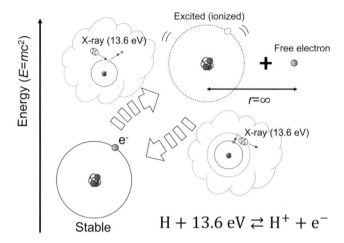

Fig. 1.6 Difference of weights (energies) between stable and excited atoms. A sum weight of excited atom and free electron is heavier than stable atom

states of a stable hydrogen atom and an ionized hydrogen atom. Let's consider the difference between these two energy states corresponding to different states of matter. When a stable hydrogen atom is given 13.6 eV of energy, which is the binding energy of the K shell, it can be ionized. On the contrary, when the electron is bound to the proton, it means that the energy of 13.6 eV is released. From these facts, it is seen that the sum of the mass of the ionized atom and the free electron is heavier than the mass of the stable atom in which the electron is bounded.

1.4   Multielectron Atom

In this book, we will describe the response characteristics of the detector change due to the emission of characteristic X-rays. In order to understand this, it is necessary to understand the energy state of electrons in an element. Therefore, we will briefly describe the energy states of multielectron atoms.

The energy states of electrons in multielectron atoms are shown in Fig. 1.7. The $n = 1$ shell, which is K shell, has one energy band, and two electrons having upper and lower spin states can exist. This is because the electrons occupy the orbit obeying the Pauli exclusion principle. Therefore, the K shell can be filled with two electrons for the $1s^2$ shell. Because the number of electrons is equal to the atomic number, the chemical properties of the K shell electrons are presented for $_1$H and $_2$He. For the L shell, there are four energy states, and therefore this shell can be filled with eight electrons; this is described as $2s^2 2p^6$. Similarly, we can understand that the M shell has eight electrons, described as $3s^2 3p^6$. The important thing is that there are large energy gaps between different main quantum numbers. The first gap is observed between the filled K shell ($1s^2$) and L shell; therefore, the $_2$He atom has a closed shell and is a noble gas. The second gap is presented between the filled L

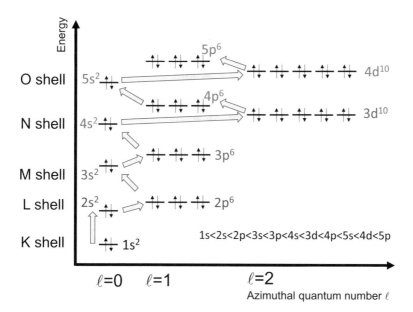

Fig. 1.7 Energy states of electrons bounded in orbitals

shell ($1s^2 2s^2 2p^6$) and M shell; therefore, the $_{10}$Ne atom has a closed shell and is also a noble gas. Similarly, $_{18}$Ar is a noble gas which corresponds to the shell gap between the M shell and N shell. From this basic idea, it is possible to easily understand the relationship between the atomic number and the number of electrons until the $4s^2$, which is $_{40}$Ca. An interesting phenomenon is that the energy level of the 3d orbital is lower than that of the 4p orbital, and similar relationships appear between the 4d and 5p orbitals.

The outermost electron of the element plays a very important role in determining chemical properties. On the other hand, the binding energy of the outermost electron is also an important factor from the viewpoint of radiation measurement. This is because the binding energy of the outermost electron is an elementary factor that determines the ease of ionization when ionizing materials using various radiation. Figure 1.8 shows the relationship between the first ionization energies and atomic numbers. Chemistry textbooks focused attention on the differences in first ionization energies between noble gases and other atoms, but it can be regarded that the first excitation energies for all atoms are presented in the range of approximately 10^{-3} to 3×10^{-2} keV. Since the energy of radiation covered in this book is at least on the order of tens of keV, which is much larger than the first excitation energies, it can be seen that differences in the chemical properties of elements do not have a significant effect on ease of ionization. Figure 1.8 also shows the relationship between K-shell excitation energies and the atomic number. It is clearly seen that the K-shell energies smoothly increase as the atomic number increases. Furthermore, this figure also represents the relationship between L_1-shell excitation energies and atomic number. It is clearly seen that the energy of the K-shell electron shows a monotonically increasing relationship with atomic number, but the L_1-shell electron

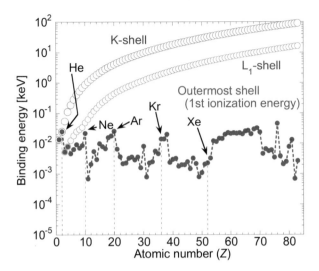

Fig. 1.8 Binding energies of K and L shell electrons (blue and green) and first ionization energy of outermost shell. The energy of K shell is monotone increasing, but first ionization energy is around 10^{-3} keV to 3×10^{-2} keV and reflected in the shell structure

has a region that is not monotonically smooth in relation to atomic numbers especially for small atomic numbers. This is because the energy of the L shell is strongly influenced by the existence of inner shell electrons. In this book, we will describe a method of elemental analysis of an object using X-rays, but the fact that the energy of the K-shell has a very good correlation with atomic number is an important tip for analyzing elements. In other words, valuable information for elemental analysis is hidden in the characteristic X-rays related to the K-shell of an element and the photoelectric effect that strongly reflects the interaction of K shell electrons.

Considering the chemical properties of elements, a table in which elements with similar properties are arranged in the same column was proposed by Mendeleev, and is known as the periodic table. Figure 1.9 shows the Periodic Table of the Elements, which is plotted along with a conceptual diagram showing the size of the elements. Radius data were determined using the following references: calculated data [5] and empirical data [6]. It is clearly seen that the atoms located in the upper right are relatively small, while the atoms located in the lower left are relatively large. There is a certain regularity, that as the size of an element decreases the atomic number increases, but this tendency may only be applied to elements within the same period, and for the elements with different periods having different main quantum numbers, it is difficult to make a general comparison.

A series of discussions revealed that the properties attributed to the outer shell electrons are less likely to have a smoothing effect with the atomic number Z, while the properties attributed to the inner shell electrons are likely to have a smoothing effect with the atomic number Z. It is possible to derive a good relationship when we focused on the physical phenomena related to inner shell electrons. As described

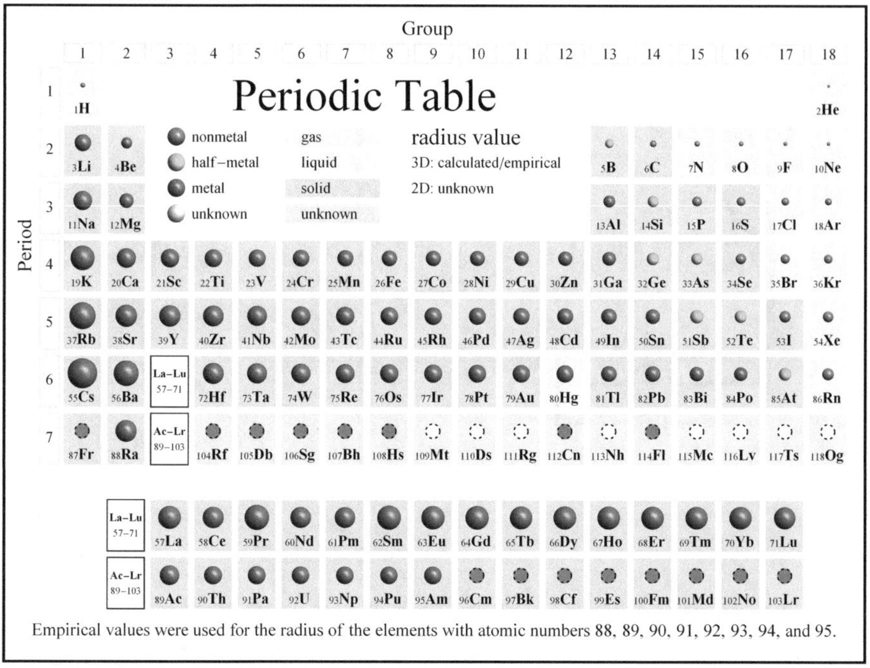

Fig. 1.9 Pediatric table with a parameter of size of each atom. The atoms located in the upper right are relatively small, while the atoms located in the lower left are relatively large

later, information concerning X-ray attenuation caused by the photoelectric effect, in which inner shell electrons play an important role, is available to derive an effective atomic number of the object.

1.5 Historical Experiment to Verify the Particle Nature of Photons

In this section, we will explain Lenard's experiment. His experiment succeeded in showing the basic experimental fact of the photoelectric effect, and gave us a very useful tool that is used in physics. Lenard revealed an important finding for explaining the particle nature of photons. Figure 1.10a is an outline of this experiment showing the experimental arrangement. A metallic plate is exposed to the light. Because this metal includes an electric circuit, this device can measure the quantity and energy of electrons generated by light irradiation. The typical result of the electric current measured with ammeter is shown in Fig. 1.10b as a function of voltage which was supplied between the metal plate and electrode; this graph was taken under the condition of using constant light frequency. The current was saturated to a constant value by increasing the supplying voltage, but this fact means that a fixed number of

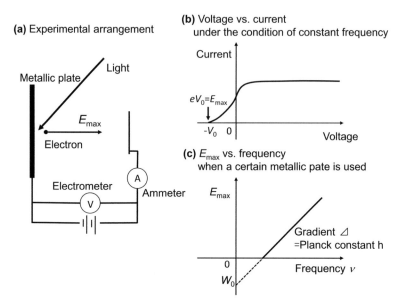

Fig. 1.10 A concept of Lenard's experiment. (**a**) shows a schematic drawing of the experiment. (**b**) shows the result of the relationship between voltage and current under the condition of constant frequency. (**c**) shows the result of the relationship between E_{max} and frequency when a metallic plate is used

electrons are generated by light irradiation. It should be noted that a certain amount of current existed even when the voltage was negative, which is the direction to cancel out the kinetic energy of the electron that was applied in the circuit. This fact means that the electron has a large amount of kinetic energy when irradiated with light. The maximum energy of the electron can be determined from the V_0 value at which the current value becomes 0. Figure 1.10c is a graph comparing the maximum energy of electrons and the frequency of irradiated light; the frequency of light changed when using the same metal plate. It was found that the maximum electron energy E_{max} has a linear relation against the frequency ν. From the graph, the following equation can be derived: $E_{max} = h\nu - W_0$, where h and W_0 are Planck constant and work function, respectively. The work function is a value that indicates the likelihood of the photoelectric effect, and is depended on the metal plate.

It was very difficult to understand the above experimental results using only classical mechanics and electromagnetics, because it was considered that the light shows wave properties from the knowledge at the beginning of the twentieth century. Einstein presented a solution to this problem, and it became possible to understand the phenomenon of the experiment by considering light as a particle having an energy $h\nu$. This experiment became known as the photoelectric effect and was interpreted as showing the particle nature of photons. Details are shown in Fig. 1.11. The left schematic drawing in Fig. 1.11 is a conceptual diagram of the photoelectric effect. The photon energy is denoted as $h\nu$, and the electron is bound in the material with binding energy "W_0". Understanding the photoelectric effect as an interaction between one photon and one electron in the material, the maximum energy "E"

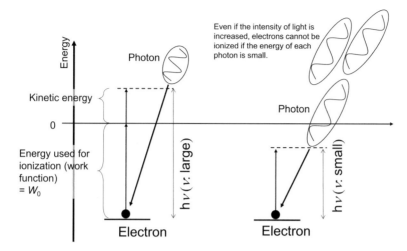

Fig. 1.11 The comparison of two situations for explaining the photoelectric effect. The left case shows X-ray incidence when one photon having high energy interacts with the bounded electron. On the other hand, shown in the right case, the photoelectric effect cannot occur even if the total energy of three photons is larger than binding energy W_0, because the photoelectric effect is the interaction between one photon and an electron bounded in the atom

given to the electron can be calculated as $E = h\nu - W_0$, and this mathematical formula is consistent with the experimental results (Fig. 1.10c). The important thing is that only one photon can interact with one electron, and this description is derived from the collision process between particles. In other words, the particle nature of the photon can be explained by this experiment. Assuming that three photons can excite one electron as illustrated in the right schematic drawing of Fig. 1.11, the electron can be excited by receiving the sum energies of the three photon's, which are lower than W_0. However, such a phenomenon has not been observed in the experiment.

To summarize the above properties, we must understand that light has both wave and particle natures. The image of light is illustrated in Fig. 1.12. Light has a minimum unit called a "photon," and its energy is uniquely determined by frequency. On the other hand, the wave nature of light is manifested by the superposition of a lot of photons. In this way, we can understand a photon, which has the properties of individual light particles, and electromagnetic wave, which has the properties of waves.

1.6   Generation of X-rays

1.6.1   *Interactions between Energetic Electrons and Target Material*

In this section, we will explain the procedure to generate X-rays and its theoretical description. Figure 1.13 is a schematic drawing of the X-ray tube and the electric circuit around the tube. X-ray imaging equipment can be controlled by the

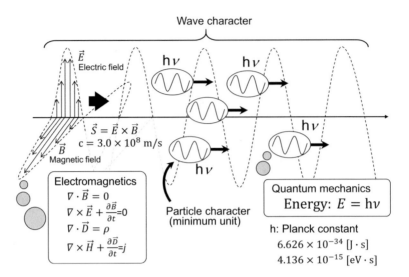

Fig. 1.12 Image of a photon and an electromagnetic wave. The photon is an individual element having an energy, hν. It can be considered that an electromagnetic wave is composed of a group of photons

following three parameters: tube voltage V [kV], electric current I [mA], and exposure time S [s]. When a current is applied to a closed-circuit that includes a filament as seen on the right side of the figure, thermoelectrons are generated from the filament. Because the filament is placed in an evacuated glass tube, the thermoelectrons are emitted into the vacuum and become free electrons. These electrons are accelerated by an electric field applied between the filament and the X-ray target, and have the energy of eV [J], also known as V [eV]. Then, by irradiating the target with energetic electrons, X-rays are emitted from the target. The detailed mechanism of how to generate X-rays from the viewpoint of interactions between electrons and target atoms will be explained later.

Figure 1.14 shows photographs of the X-ray tube. Figure 1.14a shows the exterior appearance of medical X-ray equipment which is used for general X-ray examination. Figure 1.14b is a photograph of the X-ray tube. The inside of the glass tube is evacuated and the structure of the rotating target and filament can be seen. In physics textbooks, it is sometimes shown as a schematic drawing in which the electrons emitted from the filament are uniformly irradiated to the target, but as we can see from this photograph, pinpoint irradiation used at the end of the target forms a small focus point. As a result, it is possible to irradiate an annular region in the disk-shaped rotating target with electrons. Because accelerated electrons have very high energy, a rotating target is used to avoid melting the target with heat. Figure 1.14c is an enlarged view to see the positional relationship between the filaments and the target. Most medical X-ray equipment has two types of filaments, and we can identify the large filament that can carry large currents and a small filament that can provide only small currents. When using a small filament, a small focal point can be

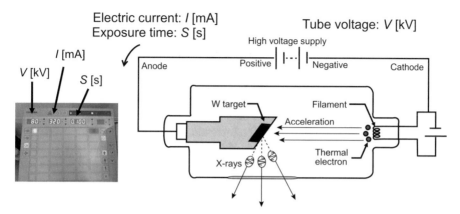

Fig. 1.13 A schematic drawing of an X-ray tube. In the electric circuit, thermal electrons were accelerated and irradiated to the target, which is generally composed of tungsten W. As shown in the photograph, tube voltage V [kV], electric current I [mA], and exposure time S [s] are important parameters. The total amount of X-rays is proportional to the product of I and S; therefore, the tube current-time product is usually called the mAs value

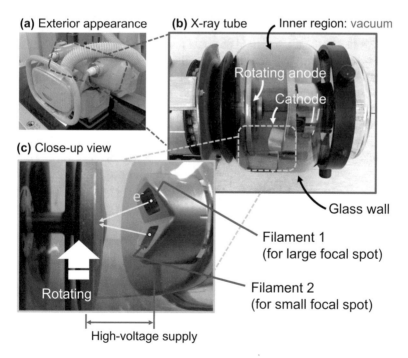

Fig. 1.14 Photographs of an X-ray tube. If we remove (**a**) the exterior appearance, (**b**) the X-ray tube can be seen. For medical use, the X-ray tube with (**c**) a rotational target is often used

formed to obtain very high image quality, but there is a demerit in that a large amount of X-rays cannot be generated at one time and the exposure time becomes long. Although it is clinically predetermined which focus should be used, many medical devices are designed to automatically select the appropriate focus, and its operation must be carried out carefully.

The electrons irradiated on the target interact with the atoms in the target material. Figure 1.15 summarizes the details of the interactions. There are three kinds of interactions: elastic scattering, collision loss, and radiation loss. "Elastic scattering" is scattering that changes the direction of incident electrons without losing energy. Collision loss is the interaction between an electron bound in an atom and an incident electron. There are two energy transfer processes depending on the atomic state resulting from the interaction. The first process is "ionization," which is the phenomenon of giving sufficient energy to the bounded electron to ionize the atoms. The electron emitted at this time has a large amount of energy and can excite other atoms, therefore the emitted electron is called a "δ-rays." "Excitation" is the process of giving the bound electron the energy that can change its orbit. The learning point is that a vacancy occurs in the inner shell during both ionization and excitation processes. The interaction described at the bottom of the figure is the "radiation loss," in which a bremsstrahlung X-ray is emitted by the interaction between the electric field of the nucleus and the incident electron. This radiation loss is the main process that emits X-rays by the electrons irradiated at the target, and we will explain this in detail later. It should be noted that bremsstrahlung X-ray emission is a relatively rare event. When an energetic electron is irradiated to the target, the energy of the incident electron is transferred into the target material through the process of elastic scattering and collision loss. As a result, diffusion of these energies occurs and the target exudes heat. This heating process occupies 99% of the interaction results, and bremsstrahlung emission occupies approximately 1%.

Two X-ray emission processes are known. One is a characteristic X-ray and the other is bremsstrahlung X-rays. We will explain the details in the following sections.

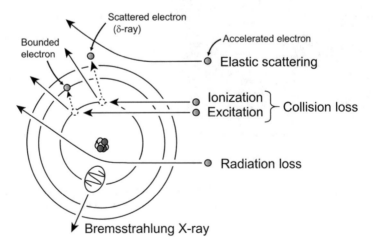

Fig. 1.15 Interaction between incident electrons and the atom. The main interactions occur with electrons bounded in the electron orbitals

1.6.2   Characteristic X-rays

Characteristic X-rays are formed in the X-ray emission process, and provided us with experimental facts that lead to the elucidation of atomic structure in the early stage of radiation research as shown in Fig. 1.2. Bohr's atomic model revealed the atomic structure which has a nucleus, and it became possible to understand characteristic X-ray emission.

Figure 1.16 is a schematic drawing used to explain the generation of characteristic X-rays. X-rays were defined as photons emitted from atoms. In order to emit X-rays, the atoms must be in an excited state. This can be achieved from the interaction with a photon (photoelectric effect) and/or an energetic electron. The electron in the inner shell is excited and this leads to the emission of characteristic X-rays. The situation in which electrons are in the excited state and there is a vacancy in the inner shell leads to an atom which is in an unstable energy state. An unstable atom in a high energy state tries to become a stable atom with a low energy state; it means that the re-arrangement of electrons occurs and the electrons existing in the high energy state orbital are de-excited so that the electron can be present in a low energy state orbital. When a transition is made from a high energy state orbital to a low energy state orbital, there is a surplus in energy, and this energy is emitted as characteristic X-rays. The characteristic X-ray emitted by the transition from the L shell to the K shell is called K_α, and the characteristic X-ray corresponding to the transition energy from the M shell to the K shell is called K_β.

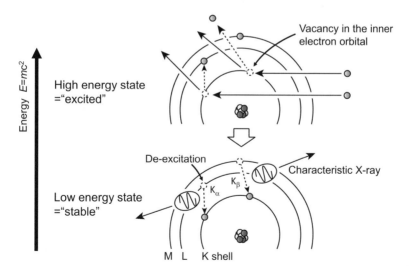

Fig. 1.16 Relationship between X-ray emission and energy states of an atom. The exciting atom is in a high energy state, and through emitting characteristic X-rays it makes a shift to a lower energy state

Most clinical X-ray equipment uses tungsten as a target material. We do so because the binding energy of the K shell of tungsten is 69.5 keV, the K-X-ray of tungsten can be observed when the tube voltage of the X-ray tube is set to 70 kV or higher. On the other hand, since the binding energy of the L shell of tungsten is 10–12 keV, it is possible that L-X-rays can be generated even when electrons are accelerated by a lower tube voltage such as 40 kV. However, since the energy of L-X-rays is also low, they are absorbed by a filter which is placed in front of the emission port of the X-rays, so they are not observed under clinical settings.

1.6.3   Polychromatic X-rays

In this section, we will describe the bremsstrahlung X-ray. Bremsstrahlung X-rays are also called polychromatic X-rays. The term "polychromatic X-ray" will be used for the explanations presented in the later sections, because a polychromatic X-ray is very clear when compared to a monochromatic X-ray. Bremsstrahlung X-rays become the main X-rays used for fluoroscopic applications, and understanding the mechanism of their generation will lead to a deeper understanding of this book, therefore we will describe the theoretical background.

Figure 1.17 shows the mechanism used to generate bremsstrahlung X-ray. Figure 1.17a shows a specific situation in which an accelerated electron passes near the nucleus and the vector of the electron is bent by the force of the electric field from the nucleus. Physical phenomena require that both energy and momentum are conserved. Since a momentum vector of the electron is determined along the electron motion direction, it is obvious that the electron momentum vectors vary significantly before and after the reaction. In the case shown in Fig. 1.17a, it can be seen that by emitting photons in the upper left direction, the reaction can satisfy the momentum conservation law. Of course, the total energy of the electron before the reaction equals the sum of the energy of the electron and the energy of the photon after the reaction, and then this reaction satisfies the energy conservation law. It is difficult to understand that bremsstrahlung radiation is continuous and the emission direction is random by only looking at this schematic drawing, so a more detailed examination will be made using Fig. 1.17b. Figure 1.17b shows the instance of electron's movement track for the case when an energetic electron is incident to the atoms arranged in a lattice. Since the electron is a charged particle, it receives Coulomb's force from every atom at every moment. As a result, the track becomes a complex trajectory that invariably changes direction, as shown in this figure. A certain interaction is shown in the upper right figure. This inset shows that the electrical force received by the electron differs depending on the positional differences between incident electrons and the nucleus, and as a result, the energy of the bremsstrahlung X-ray changes. Namely, the generation mechanism of bremsstrahlung X-rays can be quantified as presented in Fig. 1.17a, but in reality it undergoes a very complicated process as shown in Fig. 1.17b. Therefore, it is very difficult to predict the emission direction and energies of bremsstrahlung X-rays using physics.

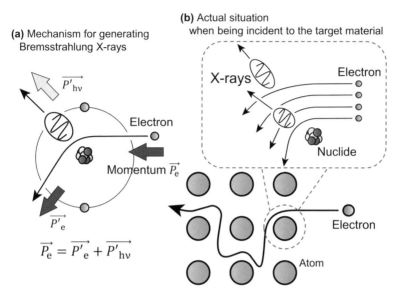

Fig. 1.17 Detailed explanations of the generation of the bremsstrahlung X-ray. (**a**) shows the concept of X-ray emission; relationship between the incident electron and generated X-ray is uniquely determined. However, in an actual situation being incident to the target material, an incident electron is randomly scattered by many atoms in the target

Although it has become possible to calculate the elementary processes of interactions step by step using computer assisted Monte-Carlo simulation, in the early stage of radiophysics research, attempts were made to reproduce X-ray spectra from an experimental approach. The following is the most famous derivation of distribution of bremsstrahlung X-rays; the method is called "Kramers' formula."

Figure 1.18 is a schematic drawing to derive the polychromatic X-ray distribution calculated using Kramers' formula. Figure 1.18a shows the energy distribution of the bremsstrahlung X-rays emitted when a thin target material is irradiated with energetic electrons. Here it is assumed that the energy is adequately high so that the electrons can penetrate the target material. When electrons with a certain energy "E_1" are incident on the thin target material, bremsstrahlung X-rays are generated. The energy intensity distribution of bremsstrahlung X-rays generated at this time is characterized as it shows constant intensity of energy 0 to energy E_1. In the mathematical description, it can be described as dI/dE = constant, where I is the intensity of X-rays. Interestingly enough, no energy dependence of dI/dE has been observed. In other words, it was determined by an experiment in which the intensity distribution of the bremsstrahlung X-ray is observed to be constant even when the incident energy varied from E_1 to E_2. This fact brings us a very important consideration. As shown in Fig. 1.17, the mechanism of the generation of bremsstrahlung X-rays has been theoretically clarified and an analytical solution can be

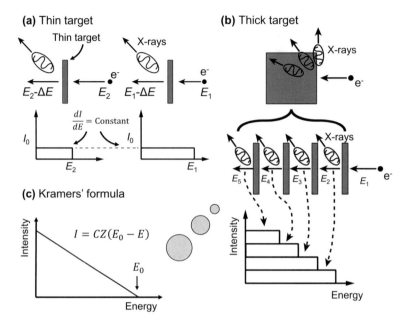

Fig. 1.18 Polychromatic X-ray distribution calculated by Kramers' formula. (**a**) and (**b**) show schematic drawings of the phenomena when energetic electrons were irradiated to target materials consisting of thin and thick targets, respectively. (**c**) represents theoretical X-ray distribution

derived for the elementary process, but it is difficult to derive bremsstrahlung X-ray distribution when energetic electrons are incident to a thick target. It should be noted that the schematic drawing shown in Fig. 1.18a is only the behavior of electrons when they are incident to a thin target, and other factors must be considered for estimating the behavior when using a thick target. Figure 1.18b shows a schematic drawing to derive the distribution function of bremsstrahlung X-rays when electrons are incident to a thick target material. Here, we assume that the thick target material is composed of the sum of thin target materials. That is, the energy E_1 loses a small amount of energy ΔE and becomes energy $E_2 = E_1 - \Delta E$, and energy E_2 also loses a small amount of energy ΔE and becomes energy E_3. At this time, it is known that ΔE takes a constant value which is not dependent on the incident energy when the electron energy is sufficiently high. Combining these considerations, as shown in Fig. 1.18b, when an electron with a certain energy E_1 is incident on a thick target material, the intensity distribution of the bremsstrahlung X-rays can be obtained by a superposition of the contributions of the bremsstrahlung X-rays related to E_2, E_3, E_4, etc. As a result, it is not difficult to imagine that the distribution function will be like a staircase, as shown at the bottom of Fig. 1.18b. This result can be obtained when we set a large ΔE, and in order to estimate realistic X-ray distribution, we should make ΔE small. Under this consideration, we can understand that the X-ray distribution becomes a linear function as shown in

Fig. 1.18c. This is called Kramers' formula. The feature of this formula is that the endpoint (maximum energy: E_0) of the linear function becomes an energy equivalent to the tube voltage.

1.6.4   Attenuation of X-rays

In this section, we describe the macroscopic nature of X-rays. Although we will explain the detail of the interaction between X-rays and matter in the later sections, we can understand the overall behavior of X-rays without knowledge of the interactions. This phenomenon is called attenuation.

Figure 1.19 shows a schematic drawing when X-rays enter an object. Here, we assume a situation in which a number of X-rays, I_0, having monochromatic energy E is incident to the object. The thickness of the substance is defined as "x". At this time, we can observe a phenomenon in which certain X-rays interact with atoms in the substance and disappear, and other X-rays penetrate without interacting. An important point to understand attenuation is that X-rays that pass through without interaction have energy E, which has exactly the same energy as the incident X-rays. In other words, the phenomenon of attenuation means that the number of incident photons decreases without losing their energy.

Next, we will describe the phenomenon of attenuation based on mathematics as shown in Fig. 1.19. When dI was described as the reduction rate of X-rays for penetrating the thickness dx, a differential equation to describe attenuation can be expressed as $dI/dx = -\mu I$. μ is a constant called the "linear attenuation coefficient," which is a unique value for a substance and is a function of energy. The physical implications of this constant indicate the ease of interaction. This equation shows that the changing amount of X-rays (dI/dx) is proportional to the product of linear attenuation coefficient μ and the number of incident X-ray I. An analytical solution of intensity "I" can be derived by solving this differential equation. Namely, that is expressed as

$$I = I_0 \exp(-\mu x), \tag{1.1}$$

where I_0 represents an initial value of I. This expression can also be described as

$$I = I_0 \left(\frac{1}{2} \right)^{\frac{x}{\text{HVL}}}, \tag{1.2}$$

where HVL is the half-value layer which is the thickness of the substance when the intensity of the penetrating X-ray is one half. HVL can also be described using μ, and the relationship is HVL $= \text{Ln}(2)/\mu$. The graph shown in the lower right of Fig. 1.19 is a plot of the X-ray penetration intensity as a function of the material

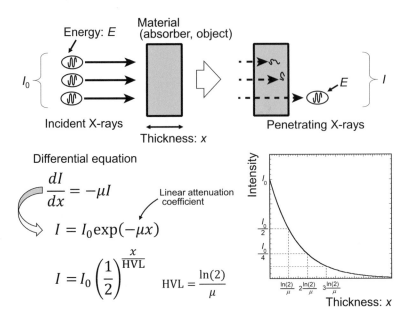

Fig. 1.19 Brief explanation of interactions between X-rays and material. We should understand that from attenuation which is based on the differential equation, a famous relationship can be derived. Note that in the attenuation process energies of penetrating X-rays are not changed even though the number of X-rays is changed

thickness. It is shown that attenuation is exponential, and $I = I_0/2$ is found at the position, $x = $ HVL.

Here, we present schematic drawing Fig. 1.20 so that the attenuation phenomenon can be visualized more clearly. The upper part of Fig. 1.20 shows a conceptual diagram when many photons are incident on a substance in which the atoms have a large interaction probability. The interaction probability is expressed by the linear attenuation coefficient μ. Since it is described by the dimension of the cross-sections, we can imagine the conceptual atomic size as shown in the figure. The rate of interaction is proportional to the product of intensity I of the incident photons and the linear attenuation coefficient μ. From this we can understand that the example shown in the upper figure of Fig. 1.20 represents a large X-ray attenuation. On the other hand, in the example shown in the lower part of Fig. 1.20, we can understand the situation in which X-ray attenuation hardly occurs. In this case, a small number of photons are incident to an atom having a small linear attenuation coefficient. The actual size of an atom can be determined by the diameter of the orbit of the outermost shell electron, however, we should note that the conceptual size discussed in this section is based on the consideration of interactions.

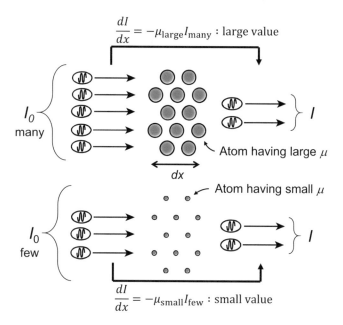

Fig. 1.20 A concept of calculating the attenuation of X-rays. In the upper case, a large attenuation is observed because many X-rays are incident to the object having larger linear attenuation coefficient. On the other hand, in the lower case a small attenuation is observed

1.7 X-ray Spectrum

We have already explained the distribution function of continuous X-rays, but the X-ray distribution (spectrum) used in an actual clinical examination is different. In this chapter, we will explain in more detail the distribution of X-rays generated using clinical equipment.

When we want to use X-rays in a clinical situation, it is very convenient if the X-ray irradiation range can be visualized. This is because we cannot see X-rays with our eyes. One solution is to install an additional device that indicates the irradiation field of X-rays. Fluorescent substances, which emit visible light when X-rays are irradiating, can be used, but it is essential to be able to identify the irradiation area without radiation. A device using visible light has been used with actual clinical equipment, and such a device is called a "movable diaphragm." Figure 1.21 shows a cross-sectional view of the X-ray emission port of clinical X-ray equipment. In the previous chapter, we introduced the Kramers' equation to obtain continuous X-ray spectra. The X-ray spectrum that can be expressed by this formula corresponds to the spectrum immediately after being emitted from the X-ray target, as shown with the blue line in Fig. 1.21.

The movable diaphragm used in clinical equipment has three advantages. One is the function of visualizing the irradiation range of X-rays, and the other is the function of collimation that allows the square irradiation area to be changed freely, and the last function is filtration. In order to perform collimation of X-rays, a metal to absorb X-rays such as tungsten should be used so that X-rays are not irradiated outside the irradiation field. A visible light source and a mirror are necessary to illuminate the X-ray irradiation field with visible light. Figure 1.21 shows the positional relationship between the visible light source and the mirror. It is nice that a mirror can blend X-rays and visible light. Assuming that the mirror is installed at an angle of 45° as shown in this figure, the distance from the center of the mirror and the X-ray focal point should be the same as the distance between the center of the mirror and the visible light source. If we want to design a more compact movable aperture, it is not difficult to imagine that the mirror angle should be set at a narrow position and the light source should be placed near the irradiation window. An important factor is that in any case, it is necessary to put a mirror in the X-ray irradiation path. This fact means that X-rays are absorbed by the structure in the diaphragm apparatus. Fortunately, it is known that low energy X-rays are less useful for imaging, and it is desirable to reduce these X-rays. Appropriately, this movable diaphragm unit is also provided with the function of filtering low energy X-rays (equivalent to 2.5 mm aluminum). As a result, as shown in the lower left figure of Fig. 1.21, medical X-ray diagnosis is performed using the X-ray spectrum in which low energy X-rays are sufficiently attenuated.

We summarize the above topics, and the relationship between the mechanisms of X-ray generation and the X-ray spectrum in Fig. 1.22. Figure 1.22a and b shows schematic drawings of bremsstrahlung X-rays and characteristic X-rays, respectively, to help with the understanding the mechanisms. Figure 1.22c is the expected

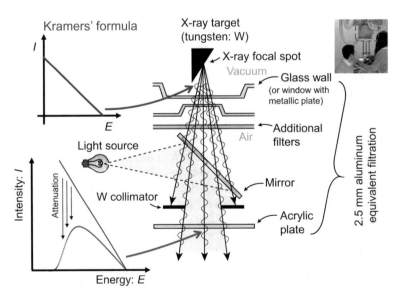

Fig. 1.21 Difference between an actual X-ray spectrum and Kramers' formula. X-ray attenuations in the movable diaphragm, which is used to form a proper irradiation area, play an important role

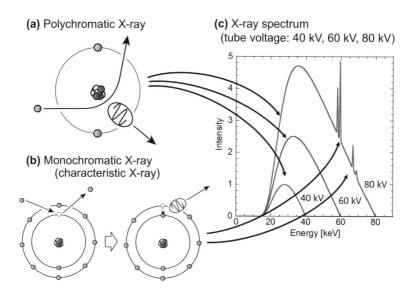

Fig. 1.22 The relationship between X-ray spectra and physical phenomena. There are two different causes for the generation of X-rays

X-ray spectrum. The figure shows the spectra of tube voltages 40 kV, 60 kV, and 80 kV, and these X-ray spectra were acquired under the condition of constant tube current. It can be seen that most of the spectrum is occupied by polychromatic X-rays, and low energy X-rays are sufficiently reduced. Furthermore, the spectrum related to the characteristic X-rays of the target material is clearly visible in the X-ray spectrum at the tube voltage of 80 kV.

Figure 1.23 summarizes the X-rays generated by the X-ray imaging system from the viewpoint of efficiency. First of all, we must note that the X-ray generation phenomenon is about 1% of the total energy amount of electrons brought into the target. From this we can see that the generation of bremsstrahlung X-rays is a very rare event, and via the interaction between incident electrons and target atoms, the incident energy is transformed into heat. Because of these circumstances, tungsten is often used as the target material, and is characterized by high atomic number and very high melting point. The part of the target irradiated with electrons becomes the point source, and X-rays generated at the target are emitted in all directions. Of these X-rays, the portion that can irradiate the subject and contributes to the creation of an image is about 1%. To understand this phenomenon, the concept of a solid angle is useful. A solid angle is defined as the surface area of a sphere with a unit radius included within the angle θ of the vertical axis, but it can be calculated using a simple function of θ. From these considerations, we can understand that only 0.01% of the energy that electrons initially bring to the target can contribute to image creation. Interestingly, since the principle of X-ray generation was discovered, no essential technological innovations have been made, and the efficiency of X-ray generation has not been dramatically improved. In addition, the X-rays

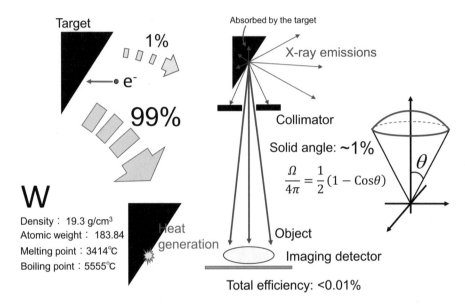

Fig. 1.23 Phenomenon succeeding to the electron irradiation to the target material. 99% of the energy is wasted as heat, and only 1% of it is used as X-ray emission. Furthermore, since X-rays are emitted in all directions, only some X-rays emitted within the solid angle are available

emitted using such mechanisms are continuous X-rays, and we have not succeeded in the creation of monochromatic X-rays to a usable level for medicine. Therefore, in this book, we wanted to propose a novel method that can maximize the features of continuous X-rays from the perspective of innovation X-ray detection.

An important factor that must be remembered when discussing the intensity distribution of X-rays is the phenomenon in which the energy distribution varies depending on the irradiation position of X-rays and this is called the "heel effect" [7]. Figure 1.24 shows a schematic drawing explaining the heel effect. First, as shown in the inset, we must note that electrons are incident perpendicularly to the target surface, and bremsstrahlung X-rays are emitted from a position which slightly penetrates the target. The generated X-rays are emitted in various directions as discussed above, but some pass through the target before irradiating the object. The X-rays emitted to the Anode side are more affected by attenuation inside the target than the X-rays emitted to the Cathode side. As a result, the effective energy of the X-ray spectrum is higher on the Anode side. Figure 1.24 shows that the X-ray spectrum changes due to the heel effect.

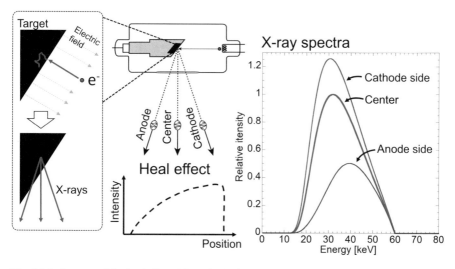

Fig. 1.24 Concept of the heel effect. Not only the intensities of the X-rays but also X-ray spectra are varied

1.8   Application of X-rays

Recently, X-rays are used in various fields. Figure 1.25 shows examples of the use of X-ray devices being applied to industry. The figure on the left shows a foreign substance inspection device, which is used to check for the presence of foreign substances in finished products using a fluoroscopic examination technique. Although high atomic number substances such as metal fragments can be identified with extremely high accuracy, other substances with similar atomic numbers such as insects and hair are similar to that of food and it is difficult to discriminate these substances based on only the X-ray attenuation rate information. Therefore, it is necessary to perform additional analysis to obtain a fluoroscopic X-ray image. The figure on the right shows a schematic drawing of baggage inspection, which is often performed for better security at airports and concert venues, etc. Deadly weapons such as knives and guns are often made with high atomic number materials. They show a large X-ray absorption so can be identified with very high precision. On the other hand, plastic bombs are relatively difficult to distinguish because they are made with low atomic number materials (except for electrical circuits). In most cases, the observational judgment of technical staff based on the shape of the fluoroscopic X-ray image is needed to determine the safety of objects.

Figure 1.26 shows an example of an X-ray generator used in medicine. Figure 1.26a is a schematic drawing of chest radiography. In a general procedure the patient is positioned so as to hold the imaging detector, and X-rays are irradiated for a short time while without breathing to obtain a sharp image without distortion [8]. Since it is better for X-rays to enter the human body in parallel line, as much as possible, we should increase the distance between the detector and/or object and the

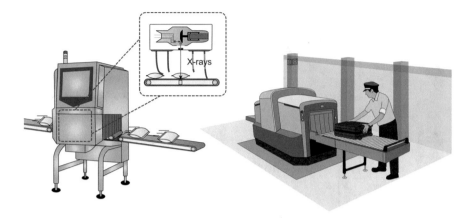

Fig. 1.25 Example of X-ray applications for industry. There is a demand for non-destructive inspection, and it is used to detect foreign substances in products

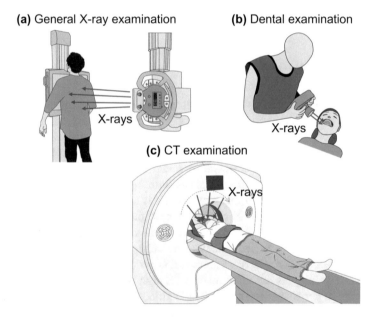

Fig. 1.26 Some examples of the use of X-rays in the medical diagnosis. (**a**), (**b**), and (**c**) show general X-ray examination, dental examination, and computed tomography (CT) examination, respectively

X-ray generator. The common distance used to take a standard X-ray is 1.5–2.0 m. The detector is a quadrangle having sides of approximately 40 cm, and X-rays are collimated so that they are appropriately irradiated within the desired area. Figure 1.26b shows an application example for dentistry. In the case of dental imaging, it is often to use a small detector that can photograph two or three teeth, and the

X-ray equipment is also placed near the patient. The irradiation X-ray field is often a few centimeters spot. Many devices do not have a movable diaphragm to identify the irradiation field with visible light, but they have a cylinder guide so that an X-ray photograph of the target object can be taken easily. Although there are special devices, such as panoramic radiography, that can acquire the information of the whole mouth at one time, in order to obtain detailed information of one tooth, a simple fluoroscopic X-ray photograph is still available for current diagnosis. Figure 1.26c is an example of computed tomography (CT) imaging. In the CT system, the patient is introduced into the equipment which has a pair of X-ray tubes and imaging detectors rotate. The internal structure of the three-dimensional human body can be measured. Detailed information can be obtained for tissue, such as bone, which attenuates X-rays appropriately, but the attenuation of X-rays for soft tissue is very similar, so the resolution to soft tissues is not always appropriate. Additionally, there is also an inspection method that uses a contrast medium (iodine) injected into blood vessels to further define soft tissue and obtain more information during a CT examination. The information obtained by CT is very useful for current diagnosis.

A very interesting point is that the polychromatic X-rays that have been described in this section generally used in X-ray examinations are now also being applied to industry and medical diagnosis. As discussed in the next chapter, the physics involved in the creation of X-ray images is rarely used in the analysis of X-ray images. If any analysis technique related to X-ray image analysis such as those described in this book can be innovated, a great spillover effect will be expected.

References

1. D.J. Holtzmann, W.T. Johnson, T.E. Southard, J.A. Khademi, P.J. Chang, E.M. Rivera, Storage-phosphor computed radiography versus film radiography in the detection of pathologic periradicular bone loss in cadavers. Oral Surg. Oral Med. Oral Pathol. Oral Radiol. Endodontol. **86**(1), 90–97 (1998). https://doi.org/10.1016/S1079-2104(98)90156-1
2. M. Sonoda, M. Takano, J. Miyahara, H. Kato, Computed radiography utilizing scanning laser stimulated luminescence. Radiology **148**(3), 833–838 (1983). https://doi.org/10.1148/radiology.148.3.6878707
3. H.G. Chotas, J.T. Dobbins, C.E. Ravin, Principles of digital radiography with large-area, electronically readable detectors: a review of the basics. Radiology **210**(3), 595–599 (1999). https://doi.org/10.1148/radiology.210.3.r99mr15595
4. T. Matsuura, H. Hayashi, H. Hanamitsu, S. Nishihara, Production of a leaf electroscope having separators and proposal of an experiment using the diagnostic X-ray equipment. Japn. J. Radiol. Technol. **69**(3), 239–243 (2012). https://doi.org/10.6009/jjrt.2013_JSRT_69.3.239
5. E. Clementi, D.L. Raimondi, W.P. Reinhardt, Atomic screening constants from SCF functions. II. Atoms with 37 to 86 electrons. J. Chem. Phys. **47**(4), 1300–1307 (1967). https://doi.org/10.1063/1.1712084
6. J.C. Slater, Atomic radii in crystals. J. Chem. Phys. **41**, 3199 (1964). https://doi.org/10.1063/1.1725697
7. S.L. Frits, W.H. Livingston, A comparison of computed and measured heel effect for various target angles. Med. Phys. **9**(2), 216–219 (1982). https://doi.org/10.1118/1.595074
8. J. Lampignano, L.E. Kendrick, *Bontrager's Textbook of Radiographic Positioning and Related Anatomy* (Mosby, Maryland Heights, 2017), pp. 1–848. ISBN-10: 9780323399661

Chapter 2
Introduction to Physics in Medical X-ray Diagnosis

2.1 X-ray Imaging

We will start this chapter with an explanation of the conventional way to understand X-ray images used for medical applications. We call a two-dimensional X-ray photograph "general X-ray examination," "plain X-ray examination," and/or more informal saying "Röntgen." Have you ever considered the principle of X-ray diagnosis used for medical applications? As you know, the principles of taking an X-ray photograph can be explained by physics; namely, the generation of X-rays, interaction between incident X-rays and objects, as well as interaction between penetrating X-rays and imaging detectors, etc. This information can help us understand the principles of X-ray acquisition. However, in general X-ray diagnosis, medical doctors do not use information pertaining to physics.

First of all, we will show you an example. Figure 2.1a shows the experimental arrangement; in this demonstration, medical X-ray equipment and a medical X-ray imaging detector were used. Figure 2.1b shows a photograph (left) and an X-ray image (right) of scissors. In this demonstration, scissors for general use and those used by children are presented. From the photograph taken by an optical camera, we cannot identify any differences, but from the X-ray image, we can identify the difference. The scissors for general use are presented in white, and those used by children are shown in gray. More clearly, we can use the following facts to understand the difference; general scissors are made of metal, and scissors made for children consist of plastic for safety. From this demonstration, it is clearly understood that an X-ray image is useful for material identification. In the above explanation using Fig. 2.1a, the most surprising thing is that we can recognize the images are of scissors when we look at only the X-ray image. At this time, we did not explain the procedure to generate the X-ray images using Fig. 2.1a. The reason why we can identify the scissors in the X-ray image is very simple; we can imagine actual scissors from the structure presented in the X-ray images. Figure 2.1c is a different

H. Hayashi et al., *Photon Counting Detectors for X-ray Imaging*,
https://doi.org/10.1007/978-3-030-62680-8_2

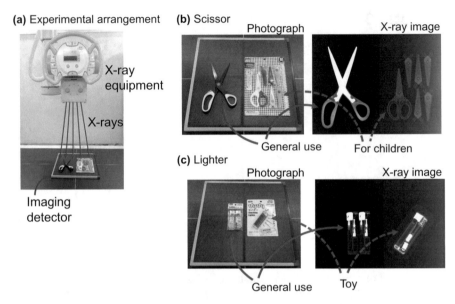

Fig. 2.1 Example of X-ray images measured with a conventional X-ray detector. (**a**) is experimental arrangement. (**b**) and (**c**) photographs and X-ray images of scissors and lighters, respectively

demonstration; two lighters are analyzed in the same manner. One is an actual lighter and the other is a toy. The differences between them are unclear when we see the X-ray images. In this case, it is difficult for us to discern the inner structure of the lighter without special knowledge. The meaning of such a simple demonstration is that X-ray images, in which the shape of an object can be easy to visualize, do not necessarily require special knowledge on radiophysics, and it can be estimated from a photograph.

Before explaining the imaging procedure to generate an X-ray image, we will explain the differences between a photograph and an X-ray image. Figure 2.2 shows a schematic drawing of an imaging technique; at this time, we will focus attention on imaging probes. As shown in Fig. 2.2a, everyone can understand the concept of the optical camera. In this case, the probe is sunlight. In a camera, the lens controls the traveling direction of visible lights, and by selecting the light that is scattered in a specific direction, an image structure of the light emitter can be obtained. Currently, we can use a two-dimensional digital sensor for photography. Interestingly, we can make an image using a small sensor, because by refracting light, an image that is upside down, with left and right reversed, can be produced on a small sensor. It is a very interesting point that the same image can be properly produced on sensors of various sizes by using the property that light bends. In the case of general single-lens reflex cameras, an image sensor with a size of 24 mm × 36 mm is often used. The number of pixels of the sensor is increasing yearly, and high-end models have tens of millions of pixels. This means that the pixel size is several microns. It is known that various colors can be expressed by the three primary colors of light of

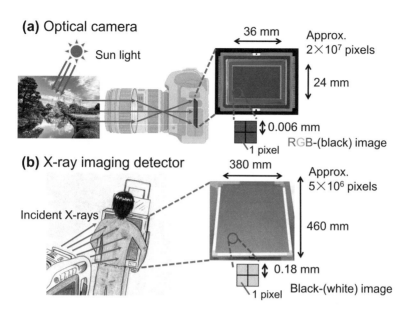

Fig. 2.2 Comparison of imaging procedures; (**a**) and (**b**) show imaging sensors of an optical camera and X-ray imaging detector, respectively

red, green, and blue. In order to imitate this method with a digital sensor, the image sensor is crafted so that the incident intensity of the three primary colors can be analyzed separately; namely, one light color can be reproduced by three light-receiving elements equipped with filters for cutting wavelengths other than the intended one. This sensor can also express brightness and darkness, that is, black and white, depending on the magnitude of the responses received.

In contrast, X-rays are not easily refracted, therefore most X-ray images are produced using X-rays which penetrate the object. Figure 2.2b shows the concept of a general X-ray examination carried out at a clinic; at this time, X-rays which are generated by the X-ray tube become the probe. Compared to the procedure used by an optical camera, we should use a relatively large two-dimensional sensor for X-ray imaging. The X-ray image is usually expressed in gray-scale. The rule for coloring the image is simple; pixels that have absorbed a large amount of X-ray energy are shown in black, and pixels that have absorbed a small amount of X-ray energy are shown in white. Figure 2.3 shows a schematic drawing of X-ray diagnosis and its basic rule to express the X-ray image. Figure 2.3a shows the concept of producing a gray-scale image. The pixel of X-rays without any absorptions in the object is expressed in black, the pixel with smaller X-ray incidence because of X-ray absorption in objects is expressed as gray or white. In order to analyze medical X-ray images, materials such as soft tissue (atomic number $Z \sim 6.5$) and bone ($Z \sim 13$) should be analyzed. As presented in Fig. 2.3a, soft tissue doesn't absorb many X-rays, and bone absorbs a great number of X-rays. Thus, soft tissue and bone are expressed as gray and white, respectively. We can imagine the situation of taking

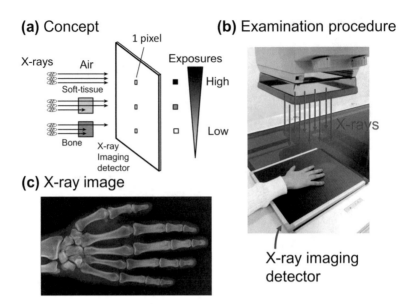

Fig. 2.3 Schematic drawing of the procedure to generate an X-ray image. (**a**) shows concept, (**b**) is experimental arrangement, and (**c**) X-ray image of a human hand

an X-ray image of the hand referring to Fig. 2.3b. Furthermore, as shown in Fig. 2.3c we can understand the X-ray image of the hand using the gray-scale rules. For clinical diagnosis using this type of image, everyone focuses their attention on bone, but from the viewpoint of physics, all of the information displayed by all pixels has important meaning. As we will describe later, when we want to analyze an effective atomic number of an object using the detection of penetrating X-rays, we also need to analyze X-rays being incident to the object; in the X-ray image, within black pixels there is information concerning the amount of incident X-rays.

2.2 Medical X-ray Diagnosis

Now, let's turn to the topic of medical X-ray imaging. Figure 2.4 shows a schematic drawing of a medical diagnosis of a chest X-ray. When we see the gray-scaled X-ray image, we can understand that the image is related to the chest region without giving a detailed explanation of the medical X-ray examination. As we explained above using scissors, a medical doctor can understand the X-ray image with the help of anatomical information of the lung region. Actually, when making a medical diagnosis, there is a lot of historical evidence in the field of medicine to support this claim, and medical doctors learn how to analyze X-ray images. Since such knowledge is shared worldwide, the level of medical diagnosis in the world is kept extremely high. In this way, the integration of knowledge for medical diagnosis is

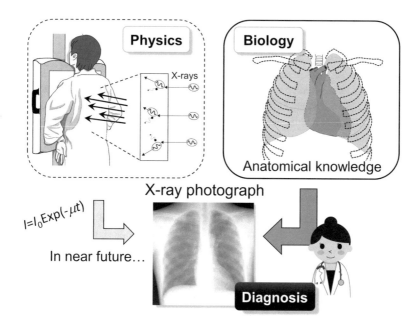

Fig. 2.4 Concept of quantitative diagnosis based on analysis of X-ray images. Currently, X-ray diagnosis generally uses gray-scaled X-ray images; diagnosis is performed using anatomy (biology). In addition to this diagnosis, we want to propose a novel diagnostic procedure which can be used to perform imaging with a photon counting detector

very useful, but we think that obsession with past knowledge should not impede the development of novel diagnostic procedures.

The biggest weakness of the current diagnostic procedure using the conventional X-ray image is that this method is not always quantitative. Namely, the result being strongly dependent on the experience of expert doctors can be a weak point in ensuring the quality of medical diagnosis. Actually, there is a problem that the diagnosis result differs depending on the medical institution. From the patient's point of view, there is also the option of visiting multiple hospitals to obtain trustworthy results, and this wastes medical resources. To improve this problem, the use of artificial intelligence (AI) for medical diagnosis is currently in the spotlight [1–3]. Although diagnosis based on AI is still under development, if an AI with a satisfactory level of skill is developed and applied to image diagnosis, the same result will be obtained regardless of which hospital we visit. This will reduce the burden on doctors and the quality of medical diagnosis is also guaranteed. However, we think that this solution alone cannot solve the problem because images used for learning by AI are based on a conventional detection principle, and it is thought that the diagnostic performance based on conventional images is limited. Research based on AI is software-based, and we think that it is necessary to conduct hardware-based research and innovation, rather than relying only on the results of AI. The diagnostic images produced by a photon counting system are images which can be used to perform quantitative diagnosis, and it is expected that photon counting can promote the improvement of diagnosis using X-ray images.

In order to understand these issues in current X-ray imaging, we will explain the basics of an imaging detector. Historically speaking, two-dimensional X-ray images can be created using detectors constructed of X-ray film, a phosphor plate, and/or a flat panel detector. It is an important point that medical doctors understand that different X-ray images are measured with different X-ray imaging detectors. Current medical imaging equipment can make a film-like image by adding image processing to the image acquired by a digital detector. This image processing plays the role of applying diagnostic evidence of past X-ray films to current digital images.

As described above, current X-ray imaging diagnosis is based on information obtained by medical doctors and radiologists, and this diagnostic method is considered to be based on anatomy. On the other hand, we know that the procedure to produce a medical image can be explained by physics. Since X-ray image diagnosis also has an application aspect based on physics, we think there is still room for developing an image analysis method which can interpret images physically and obtain quantitative diagnostic results. Thus, as illustrating in Fig. 2.4, we expect that physics-based quantitative images can provide useful information that will help with current anatomical diagnosis.

2.3   Energy Integrating Detector (EID) and Energy Resolving Photon Counting Detector (ERPCD)

The imaging detector mentioned above is called "Energy Integrating Detector (EID)," in which signals are determined proportionally to the value corresponding to totally absorbed energies ($\Sigma E(i)$) where $E(i)$ is the energy of i-th X-ray being incident to the detector. A number of electrons produced in a pixel are read out simultaneously, and they are analyzed by an integrated circuit; it is important that the amount of produced electrons is proportional to the total-absorbed energies ($\Sigma E(i)$), or more clearly, the total energy of incident X-rays. Figure 2.5a shows a schematic drawing of an EID; in this case, X-rays having different energies of E_1, E_2, E_3, and E_1 are absorbed by the EID and corresponding electrons (charge clouds) are generated in a pixel. During the generating process of an image, totally absorbed energy ($\Sigma E(i)$) values are digitalized to numbers and the digital numbers are expressed as a gray-scaled image.

On the other hand, an energy-resolving photon counting detector (ERPCD) can identify each X-ray, the reason being is that this technique uses a detection circuit to analyze individual X-rays. Another important factor is that ERPCD also has the ability to analyze each "energy" of X-ray. At the initial stage of the detection, the energy signals are processed by an analog value, at the next stage these pulses are compared with the threshold values to determine which energy band they belong to, and finally an image associated with each energy bin is created. In the case presented

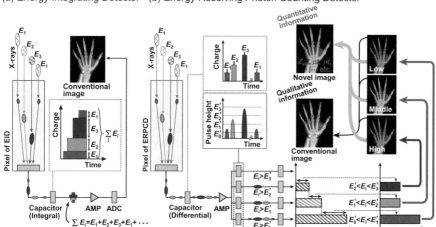

Fig. 2.5 Comparison of concept of (**a**) energy integrating detector and (**b**) energy-resolving photon counting detector. Using an energy integrating detector, the sum of energies is analyzed. On the other hand, an energy-resolving photon counting detector can analyze each energy of an X-ray

in Fig. 2.5b, a schematic drawing of a photon counting detector having three energy bins is presented: low energy bin, middle energy bin, and high energy bin. When threshold levels are set at E_0', E_1', E_2', and E_3', we can separate analogue signals into the energy regions of $E_0' < E(\mathrm{i}) < E_1'$, $E_1' < E(\mathrm{i}) < E_2'$, $E_2' < E(\mathrm{i}) < E_3'$. For the case in which there are four X-rays having energies of E_1, E_2, E_3, and E_1 being incident to the pixel of interest at different timings, E_1, E_2, and E_3 X-rays are acquired in low, middle, and high energy bins, respectively. It is important that an energy resolved image has intensity and effective energy information; therefore, we can calculate a conventional image. Based on these three images, novel analysis can be carried out. This information is completely different when compared to that of an energy integrating detector. In the later sections, we will present a procedure for using this information. Before presenting this information, we should consider the effect of the response function of a detector. We should perform this because the measured responses are not based on ideal energy absorption resulting from the interactions between incident X-rays and detector materials.

2.4   What Can We Analyze from the Images Measured with ERPCD?

When a photon counting technique is established, we can analyze the attenuation rate of the X-rays, and the analysis can lead to developing a material identification method. In this section, we want to consider the needs of material identification during medical diagnosis. A material identification procedure using X-rays is based on the analysis of an attenuation factor, which is uniquely determined by atoms. Figure 2.6a shows the mass attenuation coefficient (μ/ρ) for different atoms [4]. For medical diagnosis, analysis of soft tissue ($Z \sim 7$) and bone ($Z \sim 13$) is common practice. In addition, analysis of calcification made from calcium ($Z \sim 20$) is needed. For these atoms, the mass attenuation coefficient clearly varies compared to X-ray energy. As described later, we consider that the precise material identification in this region ($Z < 20$) can be established by measuring the μ/ρ values related to two different X-ray energies. However, the trend for mass attenuation coefficient of iodine ($Z = 53$) is different; iodine is often used in various computed tomography (CT) examinations [5, 6] such as angiography. We can see a discontinuous point at around 33 keV. This discontinuous point is called the "K-edge" meaning that attenuation of energy X-rays above 33 keV is much higher than lower energy X-rays. This phenomenon is relatively easy to identify and is applied to K-edge imaging [7].

Figure 2.6b explains the basic theory used to calculate the attenuation factor, μt. As described in Chap. 1, the attenuation of X-rays can be determined with the well-known formula:

$$I = I_0 \times \exp\left(-\mu t\right) = I_0 \times \exp\left(-\frac{\mu}{\rho} \times \rho t\right), \tag{2.1}$$

where I and I_0 are intensities of incident and penetrating X-rays, respectively. μ and t are the linear attenuation coefficient and thickness of the object, respectively. Because μ depends on the type of material, mass attenuation coefficient μ/ρ and mass thickness ρt are used instead of μ and t. When applying the formula to the analysis of ERPCD having three energy bins of low, middle, and high, we can obtain the formulas:

(a) Mass attenuation coefficient (μ/ρ)   **(b)** Procedure for calculating μt

Fig. 2.6 (a) mass attenuation coefficients for different atoms, and (b) procedure used to analyze attenuation factor from analysis of X-ray attenuation

Fig. 2.7 Effective atomic numbers of human body constituent elements

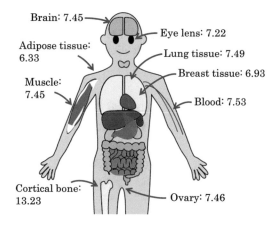

Brain: 7.45

Eye lens: 7.22

Adipose tissue: 6.33

Lung tissue: 7.49

Breast tissue: 6.93

Muscle: 7.45

Blood: 7.53

Cortical bone: 13.23

Ovary: 7.46

$$\mu_{\text{low}} t = \ln\left(\frac{I_{\text{low}}}{I'_{\text{low}}}\right), \tag{2.2a}$$

$$\mu_{\text{middle}} t = \ln\left(\frac{I_{\text{middle}}}{I'_{\text{middle}}}\right), \tag{2.2b}$$

$$\mu_{\text{high}} t = \ln\left(\frac{I_{\text{high}}}{I'_{\text{high}}}\right). \tag{2.2c}$$

In the case presented in Fig. 2.6, an attenuation factor corresponding to each energy bin can be derived. At this time, we will explain the concept of ERPCD, and under ideal conditions, the μts can be precisely determined. On the other hand, in order to analyze actual data taken with an ERPCD, we should consider the detector's response which is needed to determine μt. We will explain why our system has three energy bins, even though theoretical requirements for analysis are at most two analysis regions.

Before going to another topic, we will summarize the effective atomic numbers of the constituent atoms that make up the human body. Figure 2.7 shows a schematic drawing of the human body and the effective atomic numbers [8] of major organs. These data were calculated from the elemental composition [4] of each organ. Interestingly, most tissues made of soft tissue have very similar effective atomic numbers, which are distributed between $Z = 6$ and $Z = 7.5$. Bone is a highly absorbing substance of X-rays, therefore it is easy to image with high contrast. In fact, X-ray photographs are often used to diagnose the shape of bones. On the other hand, a detailed analysis of soft tissue is an important element in medical diagnosis, but current imaging detectors based on energy integrating can rarely perform a detailed analysis. If a photon counting type image detector is developed and has the ability to discriminate different soft tissues, it is expected to make great progress in the development of applications in the medical field.

2.5 Review of Current Novel Analysis Procedure Concerning the Analysis of Objects

In this section, we will describe current diagnostic techniques using EID. Figure 2.8a shows a schematic drawing of digital radiography using EID. In this example, we decided to use a phosphor plate as the EID. As described above, this type of detector can digitalize a value concerning the total energy deposition to one pixel during X-ray exposure. Compared with a traditional analog-type X-ray detector such as X-ray film, digital images have an advantage in that they can be utilized easily. In particular, the advantage of it being easy to copy and share data is thought to be very good for clinical work in which an electronic medical record system is generally used. It is important to note that digital detectors have become popular because of the convenience of handling data files, but their diagnostic performance has not dramatically improved.

Figure 2.8b shows another example of the use of an EID for specific precise analysis which can be applied to medical diagnosis. A "dual-energy technique" has been proposed as a device that can give quantitative information to image data

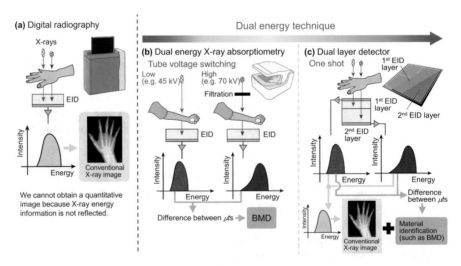

Fig. 2.8 Review of a current novel diagnostic procedure using EID. (**a**) shows generally used digital radiography using EID. (**b**) and (**c**) show dual energy techniques using EID. An example of a clinical application, analysis of bone mineral density (BMD) is one of the most important analyzing techniques

obtained by an EID. A major point is that there is a procedure using two different X-rays being generated at different tube voltages, and it is referred to as "dual-energy X-ray absorptiometry (DEXA)" [9, 10]; this technique is now being applied to analyze bone mineral density (BMD). There are some variations on how to analyze BMD using a DEXA method. In Fig. 2.8b, the schematic illustration is an example of a forearm examination. In the lower part of the figure, a schematic drawing of the X-ray spectrum is also presented. In this case, lower tube voltage X-rays (e.g., 45 kV) and higher tube voltage X-rays (e.g. 70 kV) with additional filtration are used. This technique uses the difference in μ/ρ for different effective energies of X-rays. The use of additional filtration is important to increase the difference of the effective energy of the X-rays. BMD is generally used in orthopedic diagnosis, and it is common to use DEXA equipment in medium and large hospitals.

Now, we will introduce another idea for the development of a novel X-ray imaging detector. Figure 2.8c shows a novel imaging detector being composed of two detection layers [11]; the first layer consists of a thin detector for detecting lower energy X-rays, and the second layer consists of a thick detector for detecting higher energy X-rays. The feature of this detector is that two kinds of energy information can be acquired in one imaging session with one-time exposure. At this time, a clinical trial study is underway, and we expect good results. From these examples, it can be seen that the development of energy analysis technology in the field of X-rays has high expectations for use as a clinical application.

The above explanations of current X-ray diagnostics are based on the analysis of signals measured with an EID, and the following is a specific example of using the ERPCD in actual clinical diagnostics. Figure 2.9 shows a schematic drawing of mammographic equipment in which a Si detector is used as an ERPCD. This device is preliminarily used to analyze resected samples of the breast in which there exists a large amount of cancer cells. In previous reports, the authors reported that the ERPCD can distinguish materials by means of the analysis of X-ray attenuations based on physics [12–14]. As explained in Fig. 2.5, ERPCD can derive not only a conventional image but also additional image information which can be used for

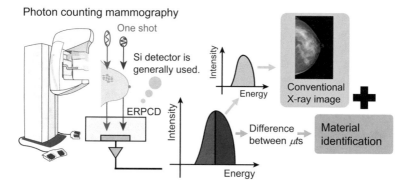

Fig. 2.9 A specific example to use ERPCD for the current X-ray diagnosis. Some procedures for mammography examination use ERPCD, which is composed of a Si detector

Fig. 2.10 Improved accuracy of material information using a dual energy technique. The dual energy scan should solve the problem of contamination, which is presented in the shaded region of the above X-ray spectrum

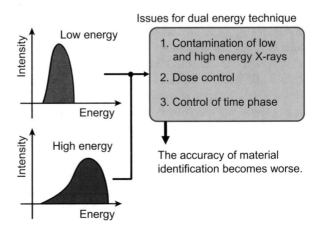

material identification based on the analysis of the differences between μts. Analysis using EID shown in Fig. 2.8b needs two X-ray exposure, but in the case of using an ERPCD as shown in Fig. 2.9, one can analyze materials using only a single X-ray exposure.

Although novel analyzing techniques based on photon-counting have been studied, it has been reported that there are some problems which reduce the accuracy of quantitative analysis [15]. Figure 2.10 explains an issue when using the dual-energy technique to derive X-ray attenuations using two different energies. In this figure, red and blue regions indicate X-ray spectra for different tube voltages. Here we should be concerned with the hatching region, which is found in the low energy region. In an ideal examination using a dual-energy technique, two different monochromatic X-rays should be used. However, for the application to medical X-ray diagnosis, medical equipment requires the use of polychromatic X-rays due to mechanical restrictions. Although high energy X-rays are usually filtered using absorbers, contamination cannot be eliminated. This contamination reduces the accuracy of the analysis. Evidence suggests that dual-energy works well in cases where there is a relatively large difference in atomic numbers, such as separation between contrast medium and calcification, but the application of its use in other clinical cases has rarely been reported. ERPCD uses a different approach, and we expect that there will be a breakthrough using an ERPCD.

2.6   Outline to Derive Precise Material Identification Using an ERPCD

In order to perform material identification through the analysis of attenuation of X-rays of an object, we should be concerned with the beam hardening effect of an object and the response function of the detector. Figure 2.11 shows an illustration of these issues. We should use polychromatic X-rays for carrying out diagnosis instead

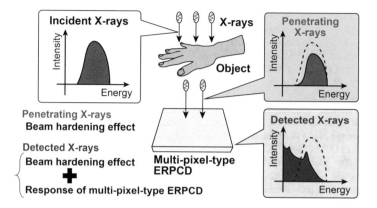

Fig. 2.11 Outline of how to perform precise material identification using an ERPCD. In order to analyze the X-ray attenuations in the object for the aim of material identification, the beam hardening effect for penetrating X-rays and the response function of the detector should be accounted for

of monochromatic X-rays. The differences in effective energies should be taken into consideration when analyzing X-ray attenuations; in a penetrating X-ray spectrum, effective energy becomes higher than that of the X-rays being incident to the object. This phenomenon is called the "beam hardening effect." Another important point when analyzing X-ray attenuations is concerning the response function of the detector. As we will describe in later sections, X-rays being incident to the pixel of interest are not completely absorbed by the detector. This means that there is a large difference between the incident and detected X-rays. These phenomena can be analyzed by physics; in other words, physics can expand the potential of a medical image.

References

1. S. Kulkarni, S. Jha, Artificial intelligence, radiology, and tuberculosis: a review. Acad. Radiol. **27**(1), 71–75 (2019). https://doi.org/10.1016/j.acra.2019.10.003
2. H. Fujita, AI-based computer-aided diagnosis (AI-CAD): the latest review to read first. Radiol. Phys. Technol. **13**, 6–19 (2020). https://doi.org/10.1007/s12194-019-00552-4
3. K. Hung, C. Montalvao, R. Tanaka, T. Kawai, M.M. Bornstein, The use and performance of artificial intelligence applications in dental and maxillofacial radiology: a systematic review. Dentomaxillofacial Radiol **49**, 20190107(22 pages) (2020). https://doi.org/10.1259/dmfr.20190107
4. J.H. Hubbell, Photon mass attenuation and energy-absorption coefficients from 1 keV to 20 MeV. Int. J. Appl. Radiat. Isot. **33**(11), 1269–1290 (1982). https://doi.org/10.1016/0020-708X(82)90248-4
5. F. Zhang, L. Yang, X. Song, Y.-N. Li, Y. Jiang, X.-H. Zhang, H.-Y. Ju, J. Wu, R.-P. Chang, Feasibility study of low tube voltage (80 kVp) coronary CT angiography combined with contrast medium reduction using iterative model reconstruction (IMR) on standard BMI patients. Br Institute Radiol **89**, 20150766(9 pages) (2015). https://doi.org/10.1259/bjr.20150766

6. K.T. Bae, Intravenous contrast medium administration and scan timing at CT: considerations and approaches. Radiology **256**(1), 32–61 (2010). https://doi.org/10.1148/radiol.10090908
7. M. Sigovan, S.S. Mohamed, D.B. Ness, J. Mitchell, J.-B. Langlois, P. Coulon, E. Roessl, I. Blevis, M. Rokni, G. Rioufol, P. Douek, L. Boussel, Feasibility of improving vascular imaging in the presence of metallic stents using spectral photon counting CT and K-edge imaging. Sci. Rep. **9**, 19850(9 pages) (2019). https://doi.org/10.1038/s41598-019-56427-6
8. F.W. Spiers, Effective atomic number and energy absorption in tissues. Br. J. Radiol. **19**(218), 52–63 (1946). https://doi.org/10.1259/0007-1285-19-218-52
9. J. Haarbo, A. Gotfredsen, C. Hassager, C. Christiansen, Validation of body composition by dual energy X-ray absorptiometry (DEXA). Clin. Physiol. **11**(4), 331–341 (1991). https://doi.org/10.1111/j.1475-097X.1991.tb00662.x
10. N.J. Fuller, M.A. Laskey, M. Elia, Assessment of the composition of major body regions by dual-energy X-ray absorptiometry (DEXA), with special reference to limb muscle mass. Clin. Physiol. **12**(3), 253–266 (1992). https://doi.org/10.1111/j.1475-097X.1992.tb00831.x
11. Fujifilm cor. (https://www.fujifilm.com/), News release: https://www.itnonline.com/content/fujifilm-unveils-two-new-digital-radiography-detectors-rsna-2018
12. E. Fredenberg, D.R. Dance, P. Willsher, E. Moa, M. von Tiedemann, K.C. Young, M.G. Wallis, Measurement of breast-tissue x-ray attenuation by spectral mammography: first results on cyst fluid. Phys. Med. Biol. **58**, 8609–8620 (2013). https://doi.org/10.1088/0031-9155/58/24/8609
13. H. Johansson, M. von Tiedemann, K. Erhard, H. Heese, H. Ding, S. Molloi, E. Fredenberg, Breast-density measurement using photon-counting spectral mammography. Med. Phys. **44**, 3579–3593 (2017). https://doi.org/10.1002/mp.12279
14. E. Fredenberg, P. Willsher, W. Moa, D.R. Dance, K.C. Young, M.G. Wallis, Measurement of breast-tissue x-ray attenuation by spectral imaging: fresh and fixed normal and malignant tissue. Phys. Med. Biol. **63**, 235003 (2018). https://doi.org/10.1088/1361-6560/aaea83
15. K. Taguchi, J.S. Iwanczyk, Vision 20/20: Single photon counting x-ray detectors in medical imaging. Med. Phys. **40**, 100901-1-19 (2013). https://doi.org/10.1118/1.4820371

Chapter 3
Radiation Detector Physics

3.1 What Is Response Function?

The key point of this chapter is to understand that the detector's responses are based on physics. The basic concept of a photon counting detector is usually explained using figures like those seen in the previous sections (see Fig. 2.5), but this information is not adequate because the description in the figures does not consider the effect of physics found within a detector.

In this section, we will describe the response function of a detector. First of all, we have to take into consideration why the response function is an important element for X-ray imaging using an ERPCD. One major point of using an ERPCD is that the attenuation of X-rays using polychromatic X-ray spectra can be analyzed (Chap. 2 [15]) [1, 2]. The information which can be derived from the X-ray spectrum is necessary for identifying different materials; therefore, there is the possibility to add the new function of X-ray imaging that can be used for medical and industrial applications. Basically speaking, in order to perform precise and accurate material identification, pure X-ray spectra should be measured without distortion [2–4]. Unfortunately, the X-ray spectra measured with a spectroscopic detector are not ideal. This is because current X-ray detectors have inadequate absorption efficiency which becomes a serious problem for multi-pixel-type imaging detectors [5–7]. This insufficiency is caused by the interaction between incident X-rays and materials which are used in the imaging detector.

Before describing our method, we should investigate a material identification method using a photon counting technique. In previous references [8–10], authors analyzed X-ray attenuations which included the detector responses, and it has been reported that material separation can be performed using actual signal differences. We think that more accurate material identification can be established by taking into consideration detector responses.

H. Hayashi et al., *Photon Counting Detectors for X-ray Imaging*,
https://doi.org/10.1007/978-3-030-62680-8_3

A measured spectrum including detector responses is known as the "response function" [11, 12]. Using Fig. 3.1, we explain the concept of the response function. Figure 3.1 explains the case in which monochromatic X-rays having energy "E" are incident to the detector; as shown in the inset, the spectrum related to incident X-rays has only one line. When X-rays are incident to the detector, various interactions occur, and charges that are generated become electrical current. Then, via digital processing of the collected charges, a measured spectrum is obtained. The spectrum reflects all of the physics contained within the detector's material, and is equivalent to the probability distribution function. The measured spectrum corresponding to monochromatic X-ray is called the "response function."

This response function is considered to consist of two major components: one is a full energy absorption event, which equals energy E, and the other is affected by insufficient absorptions. What are the factors that affect a response function? Response functions reflect detector interactions, therefore characteristics of the detector, such as pixel size, detector thickness, material composition, etc., become important parameters. That is why we will explain the relationship between response function and interaction using basic knowledge of physics in this section.

Before describing the response function of a multi-pixel-type ERPCD, we will explain the traditional analytical procedure for a single-probe-type spectrometer (photon counting detector). When researchers use a single-probe-type spectrometer, we usually analyze X-ray spectra after unfolding correction [12–14] is carried out. The concept of unfolding correction is represented in Fig. 3.2. Here, G(E): black and F(E): red represent the measured and unfolded spectra (real X-ray spectrum), respectively. The unfolding process is the analysis where G(E) is converted to F(E). The response function is represented by R(E',E): blue; here E' represents the energy of the incident monochromatic X-ray. The mathematics for the unfolding procedure

Fig. 3.1 Concept of the response function. Monochromatic X-ray having energy E is incident to the detector. The measured spectrum, which is called "response function," is different than the original spectrum

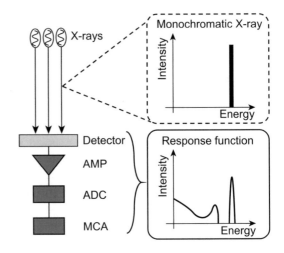

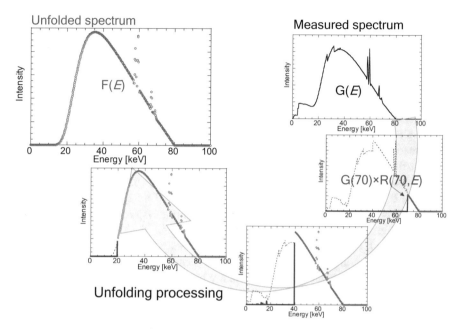

Fig. 3.2 Concept of the unfolding procedure (stripping method) of a CdTe detector. G(*E*) and F(*E*) represent measured and unfolded spectra, respectively. R(*E'*,*E*) indicates the response function of the detector; *E'* shows the energy of the incident monochromatic X-ray. It should be noted that unfolding correction can only be applied to accumulated spectra

is described elsewhere and there are many methods to achieve the unfolding procedure. The stripping method is usually used to correct continuous spectra (such as bremsstrahlung X-ray spectrum [13, 14] and beta-ray spectrum [15]). Figure 3.2 shows a typical case of the unfolding process of an X-ray spectrum (tube voltage 80 kV) measured with a single-probe-type CdTe detector. The concept of the stripping method is as follows. First, we focus attention on the intensity of the highest channel; this is a point that the highest channel does not include insufficient events. Because we know the response function concerning monochromatic X-rays found at the highest channel, we can derive the contamination rate of insufficient events caused by the highest channel. Next, we can remove (subtract) the contribution of insufficient events from the G(*E*) and the amount of insufficient events is properly corrected for the highest channel by considering interaction probability. After carrying out the above analysis, the second highest channel becomes an event without insufficient absorption. Therefore, we can iterate the above procedure to the second highest channel. In Fig. 3.2, we also present the spectrum in progression; in each figure, red circles show corrected data. The feature of this method is ease of application. However, it should be noted that there are differences in the accuracies of energy regions; high accuracy is obtained in the high energy regions and relatively low accuracy is in the low energy regions. Fortunately, because the analysis of the X-ray spectrum and the beta-ray spectrum focuses attention on the higher energy region rather than the lower region, the stripping method works well.

Although the unfolding correction procedure can be applied to derive the actual X-ray spectrum, it should be noted that the correction can only be applied to "spectrum" measured using a precise energy-resolving system under the condition of the statistical uncertainty being reduced to a sufficient level. For the conventional way to measure spectrum, we usually use a multi-channel-analyzer (MCA) having many energy bins (0.5 k to 4 k channels). In this case, unfolding correction can be applied to the accumulated spectrum. This point becomes a critical problem in the development of a multi-pixel-type ERPCD. Most of the multi-pixel-type imaging detectors produced using current technology can only perform an analysis using a few energy bins [16]. More clearly, it seems to be difficult to apply the conventional unfolding method to a measurement system using multi-pixel-type imaging detectors. It is believed that a solution will be proposed in the near future, but at present, most applications cannot use several energy bins. Even if a system with a large number of energy bins could be put into practical use, in this case, the amount of data in each bin is reduced, and additional effort to suppress statistical fluctuations is needed. Therefore, if we could develop a technology that can analyze X-ray spectrum information using a system having several energy bins, the knowledge would be a very important finding for the research of a multi-pixel-type imaging detector.

Next, we will describe the response function of a multi-pixel-type detector. Then, we should explain the creation of electron-hole pairs (charge cloud) in the detector material. To begin this description, we will assume the case in which the charge cloud is completely separated (pixel by pixel). Under this assumption, three main cases are considered as illustrated in Fig. 3.3. On the left, the charge cloud is only created in each pixel. In the middle, the charge cloud is created in the middle area of the pixels, and only a part of the charge cloud is measured in each pixel; namely, the total amount of the charges is divided into two pixels. On the right, a charge cloud is first created in a pixel of interest, and succeeding characteristic X-rays cause a "cross-talk event" between the adjacent pixels. Afterward, secondary charge clouds are created within the adjacent pixels. Although the physical phenomena are completely different for the middle and right conditions, the similarity is that the pixel of interest does not achieve full energy absorption. Only in the first case (left), a full energy peak is obtained in the response function. For the other two cases, parts of the energies are absorbed in the pixel of interest, and these events cause insufficient energy absorption. These phenomena are close to ideal, and actually in physics, the charge transport processes should be taken into consideration. Figure 3.4 illustrates the charge collecting process in monolithic detector material. Most of ERPCDs have been developed using monolithic semiconductor detector material, and each pixel is formed by contacting a separated electrode in the back of the detector material. As shown in the illustration, an electric field is formed for each pixel, and the generated charges are transported by this electric field. At this time, not all charges are collected in the corresponding pixel, because the charges are collected by an electrode via an electric field while diffusion occurs. In this book, we did not analyze the precise physical phenomena concerning diffusion, but the resulting effects will be properly taken into consideration as a particular characterization of the response function.

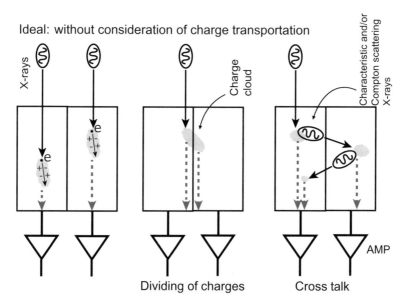

Fig. 3.3 Relationship between incident and/or scattered X-rays and detector responses. As you can see in this drawing, the presence of characteristic X-rays plays an important role in the determination of response function

Fig. 3.4 Schematic drawing of an actual situation inside a multi-pixel-type ERPCD. Electric fields and charge spread effects occur during electron transportation

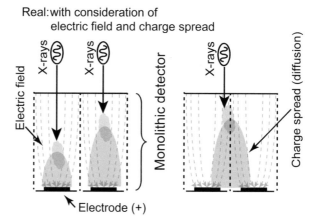

The above effects concerning insufficient energy absorptions are called the "charge sharing effect" as a collective term. There are many papers which explain charge sharing effects for test models of multi-pixel-type imaging detectors [17, 18]. These explanations go into much greater detail than the description presented in this section. Using Monte-Carlo simulation, we will mainly describe the situations represented in Fig. 3.3, and in later sections we will additionally consider the results of the charge collecting process on the measured spectra.

3.2 Interaction Provability of Semiconductor Materials Used in the ERPCD

Currently, cadmium telluride (CdTe) and/or a cadmium zinc telluride (CZT) detectors are expected to be used as photon counting detectors. In the construction of CZT detector, the ratios of Cd, Zn, and Te vary from 1-x:x:1 where x is 0.04–0.20 [11], and the contribution of Zn is much smaller than those of Cd and Te. In the following sections, we will describe the characteristics of Cd and Te as the main materials of an ERPCD.

If an ERPCD is to be used for medical examinations, it is important to use low energy X-rays as those seen in the use of Si detectors found in mammography equipment (see Fig. 2.9) (Chap. 2 [12–14]). The atomic number of Si ($Z = 14$) is relatively low, therefore the response function becomes less complicated [19]. That is why Si detectors are easier to apply to commercially available medical equipment. There are many scientific results for analyzing X-ray attenuations based on material identification using various equipment. The X-ray tube voltage for mammography examination is below 40 kV, and the detection efficiency of Si detectors maintains a high level of efficiency for detecting these X-rays. On the other hand, when using a photon counting detector for general X-ray examination, in which tube voltages between 40 and 140 kV are applied, Si detectors are not considered to be suitable.

The probability of the interactions between incident X-rays and detector material can be described as a "cross-section" [20]. Figure 3.5 shows a comparison of linear attenuation coefficients (probability of interactions) for Si and CZT. We can see that the coefficient of Si is much less than that of CZT, and in a CZT detector, the contribution of Zn is small. Although there are many studies focusing on the development of photon counting detectors using other materials such as GaAs [21], HgI_2 etc., we will only introduce CdTe and CZT detectors in the following sections. Note that it is important to obtain a good response rather than achieving high detection efficiency. Using Fig. 3.6, we will explain the importance of valances between detection efficiency and response function. As shown in the left figure, if low atomic number materials are used for detectors, the probability of interaction becomes lower. At this time, a signal sufficient to analyze the X-rays cannot be obtained. As shown in the right figure, if high atomic number materials are used in the detector, most X-rays will interact with the detector material at the surface. As we will describe later, the main interaction is the photoelectric effect, and this phenomenon produces characteristic X-ray emissions. This means that not all events result in a total absorption event. On the other hand, when middle atomic number materials are used for detector construction, those which achieve good valance for detection efficiency will give good responses. We think that CdTe and CZT have good valance properties and can be used with diagnostic X-rays.

Figure 3.7 shows cross-sections of Cd and Te atoms as a function of X-ray energy. The main interactions are coherent scattering, photoelectric effect, Compton scattering, and pair creation. Here, the cross-sections resulting from these phenom-

Fig. 3.5 Comparison of linear attenuation coefficients of CZT and Si. A CZT detector is much more efficient than a Si detector. In a CZT detector the contribution of Zn is small, because the atomic number of Zn is small and the composition ratio is also low

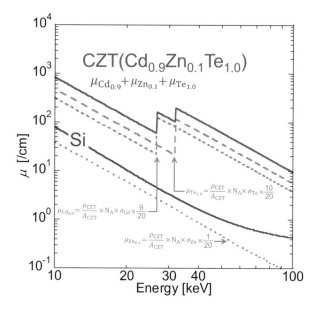

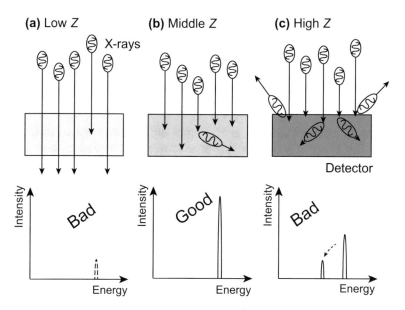

Fig. 3.6 Differences in responses for detector materials having different atomic numbers. Upper figures represent schematic drawings of a cross-sectional view, and lower figures represent expected spectra. Detector material having middle atomic numbers is proper in achieving good responses

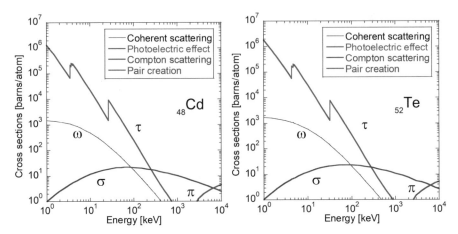

Fig. 3.7 Cross-sections of Cd and Te. Those of coherent scattering, photoelectric effect, Compton scattering, and pair creation are plotted

ena are described as ω, τ, σ, and π for coherent scattering, photoelectric effect, Compton scattering, and pair creation, respectively. Therefore, the linear attenuation coefficient μ of a certain material is described as

$$\mu = \omega + \tau + \sigma + \pi. \tag{3.1}$$

Using μ, attenuation of X-rays can be calculated using the following equation:

$$\frac{dN}{dx} = \mu N, \tag{3.2}$$

where N and x represent the number of photons and distance from the surface of the material. The equation shows the quantitative rate of the interaction in a minute range, dx. This equation demonstrates that when we know μ, the interaction rate of photons can be determined. In the sections below, we will explain these phenomena.

3.3   Interactions Between Incident X-rays and Atoms

3.3.1   Coherent Scattering

Coherent scattering is elastic scattering in which photons are treated as a wave; namely, the direction of the incident photon is changed, but the energy does not change. As shown in Fig. 3.8, there are two types of coherent scattering: one is Thomson scattering that occurs from interactions with free electrons, and the other is Rayleigh scattering that occurs from interactions with bound electrons. If coherent scattering occurs at a certain pixel in a multi-pixel-type detector, the energy of

Fig. 3.8 A schematic drawing of coherent scattering. There are two types of coherent scattering: one is Thomson scattering and the other is Rayleigh scattering

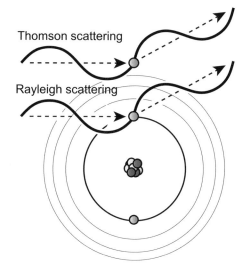

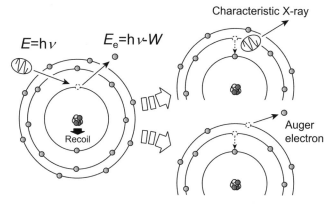

Fig. 3.9 Schematic drawings of the photoelectric effect (left), and succeeding phenomena (right). It is important to understand that additional phenomena will occur after the photoelectric effect

the photon is not absorbed in the pixel of interest. Thus, in many cases, coherent scattering does not play an important role in the estimation of the response function of a multi-pixel-type detector.

3.3.2 Photoelectric Effect

A schematic drawing of the photoelectric effect is presented in Fig. 3.9. This effect is the most important interaction when considering the response function of an ERPCD. This effect is understood as the interaction between an incident X-ray and

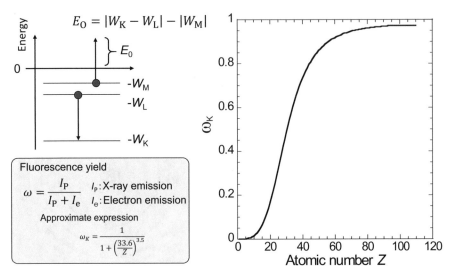

Fig. 3.10 Concept of Auger electron emission and fluorescence yield used to calculate the intensity of electron emission

an orbital electron; the orbital electron is under the condition in which the electron is strongly bound by the nucleus. Therefore, we should take into consideration the conservation laws of energy and momentum for X-rays, electron, and nucleus, when analyzing kinetics. This consideration leads to the recoil of the nuclide, but fortunately, the energy of the recoiled nuclide is much smaller than those of other particles. As a result, the energy of the recoiled nuclide is treated as negligible in most cases.

As shown in Fig. 3.9, the photoelectric effect is the interaction of an incident X-ray having energy $E(=h\nu)$ with the electron present in the shell; most importantly, the electron is not in a free condition, but bound to the atom. Then, the energy E of the X-ray is transferred to the electron which is bonded with a negative potential, $-W$. Then the electron is emitted from the atom with the release of energy, $E = h\nu - W$, and this energized electron is called a photoelectron. Fortunately, the range of generated electrons in high-dense detector material is very small, and the electron loses all of its energy within a small area. From this, a charge cloud is formed; the typical size of the charge cloud is estimated to be from $1\,\mu m$ ($E = 10\,keV$) to $15\,\mu m$ ($E = 50\,keV$) [22]. What happens next? We should note that the atom is left in an excited state after emitting the photoelectron which results in the creation of a vacancy in the orbital. Therefore, as presented in the right figure of Fig. 3.9, the remaining excitation energy is released via the emission of a characteristic X-ray or Auger electron. If the Auger electron is emitted, its energy is absorbed readily in the detector material; the resulting charge cloud may not be distinguishable from the initial cloud caused by the photoelectron. The concept of the Auger electron emission phenomenon is shown in Fig. 3.10. In the example shown in this figure, when an electron performs de-excitation from the L shell to the K shell, the corresponding energy difference is given to the M shell electron. In general, the surplus energy generated when an electron transitions to the inner shell is larger than the binding energy of the outer shell electron, therefore an electron that has received the energy has enough energy to break the bonding with the atom and be released. As already explained, when a hole is created in the inner electron shell and an electron transits from the outer shell, emission of characteristic X-rays and emission of Auger electrons compete with each other. At this time, the probability I_p through the characteristic X-ray emission process and the probability I_e through the Auger electron process can be described by the fluorescence yield ω. ω is a function of atomic number as shown in Fig. 3.10, and it shows that a smaller atomic number has a smaller ω; namely, Auger electrons are more easily emitted from smaller atomic number atoms. On the contrary, it can be seen that larger atomic number atoms have larger ω and the emission probability of characteristic X-rays becomes higher. When we want to analyze the response function of a detector from the viewpoint of physics, the trend of fluorescence yield needs to be taken into consideration. It is clearly understood that the process of emission of characteristic X-rays becomes very large when high atomic number materials are used for the detector in order to increase the cross-section of the interaction of X-rays.

The probability and intensity of characteristic X-rays are well known. Schematic drawings of characteristic X-ray emissions for Cd and Te are presented in Fig. 3.11. Cd and Te are the semiconductor detector materials used for the diagnostic X-ray region. The binding energies of Cd and Te are −26.7 keV and −31.8 keV [23], respectively, therefore the photoelectric effect can occur for X-rays having energies above these values. In this section, we limit the discussion to K-shell interaction, because the characteristic X-rays caused by the other shells are negligibly low or do not play an important role in an ERPCD used for actual medical applications. It is also important to know both the energies and intensities of K-X-rays. The lower graphs in Fig. 3.11 show distributions of K-X-rays. In this graph, intensity is defined as emission probability when 100 vacancies appear in the K-shell orbital. It is clearly seen that the intensities of K_α (de-excitation occurs from L shell to K shell) and K_β (de-excitation from the M shell to the K shell) are intense, and they account for approximately 84–86% and 5–6% of the intensities, respectively. We should take into consideration the effect of characteristic X-ray emissions when we derive the response function of a multi-pixel-type ERPCD.

3.3.3   Compton Scattering

The Compton scattering effect is the interaction between photons and free electrons. We should note that there is a difference between the photoelectric effect and Compton scattering. From a phenomenon viewpoint, the difference is clear. For the photoelectric effect, an incident X-ray disappears and succeeding characteristic X-rays are generated. On the other hand, for the Compton scattering effect, a scattered X-ray is present without the emission of characteristic X-rays. The difference

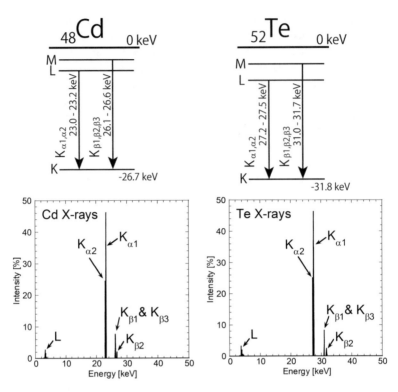

Fig. 3.11 Characteristic X-rays emitted from Te and Cd. In these calculations, intensities of X-rays are defined when 100 vacancies are present in the K-shell

of the phenomena is caused by different electron states: free electrons or bound electrons. In the situation of Compton scattering, free electrons are taken into consideration, the mathematics for conservation laws of energy and momentum between two particles should be solved. A schematic drawing of the Compton scattering effect with energy values and momentum for each particle is shown in Fig. 3.12. The energy conservation law can be described as follows:

$$h\nu = h\nu' + \text{K.E.} \tag{3.3}$$

where $h\nu$, $h\nu'$, and K.E. are the energy of incident X-ray, the energy of scattered X-ray, and kinetic energy of the scattered electron, respectively. The law of conservation of momentum can be described as follows:

$$\text{X direction}: \frac{h\nu}{c} = \frac{h\nu'}{c}\cos\phi + P_e\cos\theta, \tag{3.4a}$$

$$\text{Y direction}: 0 = -\frac{h\nu'}{c}\sin\phi + P_e\sin\theta, \tag{3.4b}$$

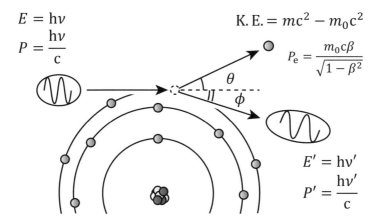

Fig. 3.12 A schematic drawing of the Compton scattering effect. The equations needed to solve Compton scattering are simple because this phenomenon occurs between an incident photon and a free electron. Using conservation laws of energy and momentum, we can calculate an analytical solution for the energy of scattering X-ray as a function of scattering angle ϕ

where $h\nu/c$, $h\nu'/c$, and P_e are momentums of incident photons, scattered photons, and scattered electrons, respectively. θ and ϕ are scattering angles of an electron and a photon, respectively. Here, we want to solve these equations for the scattered photon because measurement and analysis of an electron are often difficult. The relational expression of $\sin^2\theta + \cos^2\theta = 1$ is used to eliminate θ, and we want to solve the equations against $h\nu'$. It is an important point that the equations can be solved analytically, and the energy of scattered X-ray ($h\nu'$) can be described as

$$h\nu' = h\nu \times \frac{1}{1 + \dfrac{h\nu}{m_0 c^2}(1 - \cos\phi)}, \tag{3.5}$$

where $m_0 c^2$ is the electron at resting mass: e.g., 511 keV. Equation (3.5) indicates that the energy of a scattered X-ray is made up of incident X-ray energy $h\nu$ and scattering angle ϕ. We will show you typical examples of calculated results for 10 keV and 100 keV X-rays as presented in the upper graphs of Fig. 3.13. In this graph, the data for Compton scattering in the diagnostic X-ray region can be clearly seen; the energy of scattered X-rays is similar to that of incident X-rays at 10 keV, and the energy of scattered X-rays is slightly decreased at 100 keV. This shows that most of the energies remain as scattered X-rays when the effect occurs in material.

The above formula can be used to calculate the energy of scattered X-rays, and when considering Compton scattering, the probability of scattered angle (differential cross-section) is also very important. The mathematics are known as the Klein-Nishina formula,

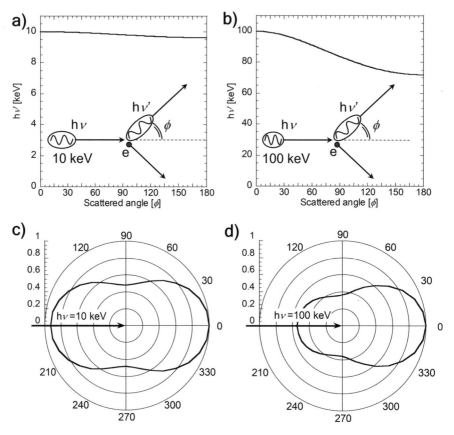

Fig. 3.13 (**a**) and (**b**) show energies for scattered X-rays from incident X-rays at 10 keV and 100 keV, respectively. (**c**) and (**d**) show intensities of scattered X-rays from incident X-rays at 10 keV and 100 keV, respectively

$$\frac{d\sigma}{d\Omega} = Zr_0^2 \left[\frac{1}{1+\alpha(1-\cos\phi)} \right]^2 \left[\frac{1+\cos^2\phi}{2} \right] \left[1 + \frac{\alpha^2(1-\cos\phi)^2}{(1+\cos^2\phi)[1+\alpha(1-\cos\phi)]} \right], \quad (3.6)$$

where α and r_0 represent $h\nu/m_0 c^2$ and the classic electron radius, respectively. α is a function of incident X-ray energy $h\nu$; therefore, this equation also contains functions of incident X-ray energy $h\nu$ and scattering angle ϕ. The relative intensities of 10 keV photons and 100 keV photons are presented as a function of angle in the lower graphs of Fig. 3.13. For 10 keV X-rays, the probabilities for forward ($\phi \sim 0°$) and backward ($\phi \sim 180°$) directions are much larger than those of side directions ($\phi \sim 90, 270°$). Although it is not always easy to understand the situation where the intensity of back-scattered X-rays is higher than those of side-scatter in classical mechanics, we should be able to understand this after looking at Fig. 3.13 which shows typical trends of quantum mechanics. On the other hand, the intensity of

100 keV X-rays for the forward direction is relatively higher. Diagnostic X-rays are in the polychromatic X-ray spectrum having energies between 15 keV and several-tens keV, and we should understand that there is a trend for higher X-rays and that for lower X-rays and their spectrums are different. That is why the response function of an ERPCD becomes complex.

3.4   Response Functions and X-ray Spectra

3.4.1   Response Function of a Single-Probe-Type CdTe Detector

Let's begin by discussing the response function of a single-probe-type CdTe detector. Figure 3.14 shows the relationship between response function and a schematic drawing of interactions of incident X-rays in a single-probe-type CdTe detector. Among these interactions, consideration of differences between a full energy absorbing event and a partial energy absorbing event is important. The main peak is caused by full energy absorption of the incident X-ray in the detector material. Namely, when all of the energies of the secondary produced particles are completely absorbed by the detector including the photoelectric effect and Compton scattering effect, a full energy peak (FEP) appears. However, it is important to consider the transportation of energies, of not only incident X-rays but also secondary produced particles. When the photoelectric effect occurs, succeeding X-rays are produced as shown above. In the case where the photoelectric effect occurs at a point deep within the detector, secondary produced X-rays may be absorbed by the detector material. On the other hand, in the case that the effect occurs at the surface of the detector material, there is the possibility that the succeeding characteristic X-rays are not absorbed by the detector material. In the latter case, the absorbed

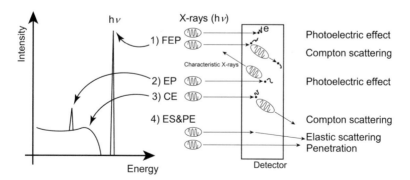

Fig. 3.14 Schematic drawing of the response function of a single-probe-type detector. Seen here are Full Energy Peak (FEP), Escape Peak (EP), Compton Escapes (CE), and Elastic Scattering and Penetration Escape (ES & PE). The corresponding phenomena are shown on the right

energy becomes slightly less than the amount of the energy corresponding to the characteristic X-rays, and this event forms peaks. These are called escape peak(s) (EP). In reality, as described above (see Fig. 3.11), the emission of the characteristic X-rays of Cd and Te is a little complex; therefore, there are many EPs corresponding to characteristic X-rays. When the Compton scattering effect occurs in the detector material, we should focus on the behavior of scattered X-rays because scattered electrons may be absorbed in the detector material. Because scattered X-ray energies are made of mostly incident X-rays (see Fig. 3.13), scattered X-rays dissipate energy. In other words, when the Compton scattering effect occurs, the detector can absorb part of the incident energies. Taking into consideration the above information, we can understand that the event corresponding to Compton scattering in the response function forms continuous areas; these are called Compton escapes (CE). The highest energy of the CE is called the "Compton edge," and it can be calculated using Eq. (3.5); the edge can be calculated as the energy of the incident X-rays ($h\nu$) minus the energy of scattered photons at a scattering angle of 180° ($h\nu'$). For 10 keV and 100 keV X-rays, they become 0.4 keV and 28 keV, respectively. This estimation is based on single events, meaning that Compton scattering occurs only one time, but in reality, multiple events may occur. The detector used for X-ray diagnosis is relatively thin, therefore multiple scattering events rarely occur. When elastic scattering and penetration events (ES&PE) occur, no events appear in the response function. However, these events should not be ignored because they lead to decreases in full energy events. From this, we can understand that the full energy peak efficiency should be considered when the unfolding process is performed on a measured spectrum.

A typical example of the response function of a single-probe-type CdTe detector is presented in Fig. 3.15. The response functions were calculated using the Monte-Carlo simulation code: EGS5 (electron-gamma-shower version 5) code [24]. Detector construction used for the simulation is presented in the inset. The detector size is 3.0 mm in width, 3.0 mm in length, and 1.0 mm in thickness; pencil and broad beams were introduced into the detection region. The picture on the right shows the actual detector modeled in the simulation. In the figure, there are three response functions corresponding to incident X-ray energies of 30 keV, 60 keV, and 80 keV; EPs and CE are clearly seen. As described above, when 30 keV X-rays are incident to the detector, the photoelectric effect only for the Cd atom can occur, because the K-shell electron of Cd is at −26.7 keV. In the response function for 30 keV photons, there can be seen a FEP and EP caused by the escape of characteristic Cd atom X-rays. It is an important point that at this energy the photoelectric effect for Te does not occur. This is because the K-shell electron is found at −31.8 keV and this value is higher than the incident energy of 30 keV. On the other hand, when the energy of incident X-rays is over 32 keV, the photoelectric effect for both Cd and Te occurs. Therefore, in the response functions at 60 keV and 80 keV in Fig. 3.15, EPs related to Cd and Te are observed. These EP energies can be calculated theoretically, and are summarized in Table 3.1. In an actual CdTe detector, the energy-resolving power is not high enough to separate all EPs as described in the table. As a result, four EPs can be seen in the response function: (1) E-Cd($K_{\alpha 1}$) and

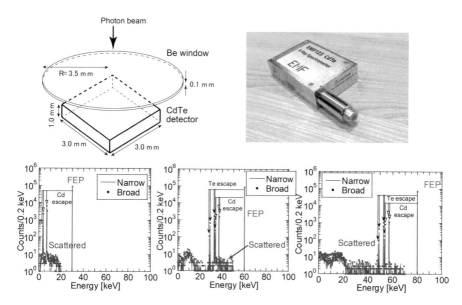

Fig. 3.15 A typical example of the response function of a single-probe-type CdTe detector. In addition to the full energy peak (FEP), we can see intense escape peaks (EP) in the response functions

Phenomena	Description	Energy [keV]
Escape (Te)	$E\text{-Te}(K_{\alpha 1})$	$E\text{-}27.5$
	$E\text{-Te}(K_{\alpha 2})$	$E\text{-}27.2$
	$E\text{-Te}(K_{\beta 1}, K_{\beta 3})$	$E\text{-}31.0$
	$E\text{-Te}(K_{\beta 2})$	$E\text{-}31.7$
Escape (Cd)	$E\text{-Cd}(K_{\alpha 1})$	$E\text{-}23.2$
	$E\text{-Cd}(K_{\alpha 2})$	$E\text{-}23.0$
	$E\text{-Cd}(K_{\beta 1}, K_{\beta 3})$	$E\text{-}26.1$

Table 3.1 Energies of the escape peaks from a single-probe-type CdTe detector

$E\text{-Cd}(K_{\alpha 2})$ at approximately 23 keV, (2) $E\text{-Cd}(K_{\beta 1}, K_{\beta 3})$ approximately at 26 keV, and (3) $E\text{-Te}(K_{\alpha 1})$ and $E\text{-Te}(K_{\alpha 2})$ at approximately 27 keV, and (4) $E\text{-Te}(K_{\beta 1}, K_{\beta 3})$ and $E\text{-Te}(K_{\beta 2})$ at approximately 31 keV. We also would like to point out the difference between narrow and broad beams. For simulations as presented in Fig. 3.15, we simulated two X-ray beam conditions: one is a pencil beam which irradiates the detector only in the center, and the other is a broad beam which irradiates the entire front surface of the detector. Fortunately, we could not identify any large differences in observed response functions as seen in Fig. 3.15. This may be due to the structure of the detector, of which the size is much larger than the mean free path of generated characteristic X-rays. As described later, when we consider a multi-pixel-type ERPCD, the size of the irradiation area becomes an important factor for determining response functions.

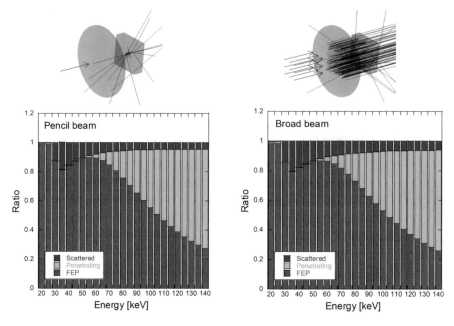

Fig. 3.16 The ratio of phenomena for the response functions of a single-probe-type CdTe detector. Although the FEP component is important for detector specification, those of scattered and penetrating are also important when all regions of the measured spectrum are used for analysis

Let's analyze the response function in detail. Figure 3.16 shows the contributing ratios of the response functions of a single-probe-type CdTe detector. Using the response functions, we analyzed the ratio of FEP, scattering parts including EP and CE, and area of penetration including ES and PE. The left and right figures demonstrate the results for pencil and broad beams, respectively. When focusing on the dependence of FEP, we can clearly find a significant unusual trend; the FEP rapidly decreases at 35 keV and gradually increases at 55 keV, and then rapidly decreases again. These phenomena are mainly caused by the photoelectric effects of Cd and Te. Please recall the cross-sections of Cd and Te in Fig. 3.7, and we can find that the trend for FEP in Fig. 3.16 is similar to the cross-section of the photoelectric effect. For energies below 60 keV which is close to the effective energy used in medical X-ray diagnosis, the ratio of FEP is approximately 80%, but we should be concerned with the other parts which account for the remaining 20%. When we want to analyze the X-ray spectrum, the effect of scattering X-rays on the response function should be considered.

A CdTe detector is convenient to use for X-ray spectroscopy, because it can operate at room temperature with high energy resolution. In addition, detection efficiency is not bad because it uses high atomic number materials ($_{48}$Cd and $_{52}$Te), however this leads to the appearance of EPs for lower energy X-rays. Therefore, the X-ray spectra measured with a single-probe-type CdTe detector needs the application of unfolding correction as described in Fig. 3.2. Figure 3.17 shows a

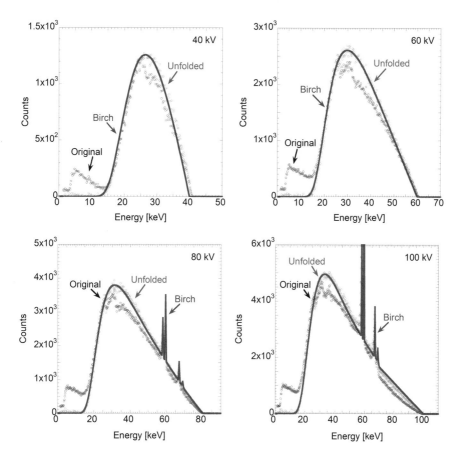

Fig. 3.17 Comparison of X-ray spectra measured with a single-probe-type CdTe detector. The measured spectrum (original spectrum) is distorted by the response function of the detector, therefore we should apply unfolding correction to obtain the X-ray spectrum

polychromatic X-ray spectra measured with a single-probe type CdTe detector. The red data points were obtained by performing the unfolding process on the measured black circle data points. Comparing the original and unfolded X-ray spectra, it can be seen that the two spectra match in the region between 15 keV and 26 keV. For X-rays between 15 keV and 26 keV the photoelectric effect in the K shell of Cd and Te does not occur, so the photoelectric effect in the L shell becomes the main contributor, and as a result, the detection efficiency is almost 100%. On the other hand, in the region above 27 keV, the unfolded spectrum has higher counts when compared with the original spectrum. The reason for this is that the phenomenon in which the FEP count is decreased due to the EP being properly corrected by the unfolding process. The binding energies of the K shells of Cd and Te are −26.7 keV and − 31.8 keV. This can be also understood from the spectrum shown in Fig. 3.17.

Namely, at the points of 26.7 keV and 31.8 keV, the characteristics of interaction and the efficiency of FEP vary greatly, and the original spectrum shows discontinuous behavior. Although utilizing these discontinuity points toward scientific analysis has not been proposed as of yet, actually, it can be used to check energy calibration and to check the operation of the unfolding program. In other words, by checking whether the 26.7 keV and 31.8 keV data points are smoothly connected by carrying out the unfolding correction, it is possible to judge whether the program is working correctly. Another important point is that the original spectrum also shows an intensity distribution in the region below 15 keV. This is not electric noise, and they are formed by superimposing the EPs continuously from high energy areas. As a check of the unfolding correction, it is important to evaluate whether this is correctly reduced. Actually, the unfolded spectrum shown in Fig. 3.17 agrees well with the Birch's formula [25], which is a semi-empirical equation used to calculate the X-ray spectra, for all tube voltages. It can be seen that the unfolding correction is performed correctly.

3.4.2   Response Function of a Multi-Pixel-Type CZT Detector

In this section, the response function of a CZT detector is described. Here, we define the response function as a spectrum having just one pixel in a monolithic detector. An illustration of the detector is presented in Fig. 3.18. The upper left and right figures show response functions for one pixel (200 µm) in a monolithic CZT detector with two different irradiation areas (side length = L), 200 µm and 400 µm, respectively. We can see obvious differences; the left graph is similar to that of the response function of the single-probe-type CdTe detector as shown in Fig. 3.15, but the right graph is different. What happens to these response functions? We will explain this phenomenon using physics.

The most important difference between the single-probe and multi-pixel detector response functions is that the multi-pixel type response function requires an additional consideration which is related to the interaction between adjacent pixels. In actual cases, multi-pixel-type ERPCDs are used in situations in which adequately large X-ray beams are irradiated to the detector, therefore we should understand the response functions of the right figure of Fig. 3.18. In the response function, we can identify the characteristic Cd and Te X-ray peaks. As shown in the left figure of Fig. 3.19, when X-rays are introduced to adjacent pixels and characteristic X-rays are produced via the photoelectric effect and there is the possibility that characteristic X-rays may enter the pixel of interest. If this occurs, characteristic X-ray peaks related to Te and Cd can be observed in the response function. Note that this phenomenon occurs uniquely in multi-pixel-type ERPCDs.

In addition, there is another difference related to characteristic X-ray emission. Remember that the binding energies of Cd and Te are −26.7 keV and − 31.8 keV, respectively, and K-X-rays of Te are from 27.2–27.5 keV; namely, the energy of K-X-rays of Te is a little bit higher than the binding energy of Cd. This means that

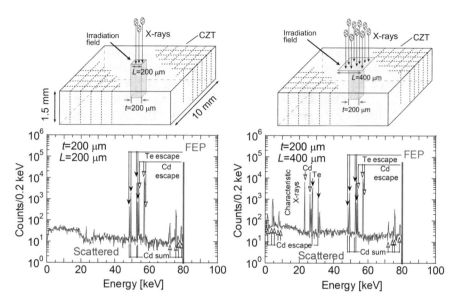

Fig. 3.18 Differences in response functions for a very small irradiation field (left: $L = 200$ μm) and a larger one (right: $L = 400$ μm) when a multi-pixel-type CZT detector was used as an ERPCD. When a large irradiation field was used, we can observe additional X-ray peaks in the energy region between 23–32 keV

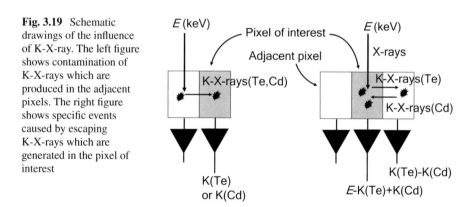

Fig. 3.19 Schematic drawings of the influence of K-X-ray. The left figure shows contamination of K-X-rays which are produced in the adjacent pixels. The right figure shows specific events caused by escaping K-X-rays which are generated in the pixel of interest

the K-X-rays of Te can interact with Cd within the detector. When this phenomenon occurs in the same pixel and/or the same single-probe-type detector, it leads to no particular effect. On the other hand, when this phenomenon occurs in different pixels, especially adjacent pixels, additional consideration is needed. The right illustration in Fig. 3.19 shows a schematic drawing of the interactions between the K-X-rays of Te and Cd. When the X-rays are incident to a pixel of interest, and the photoelectric effect occurs with Te and K-X-rays are emitted. Then, the X-rays can interact with the Cd in neighboring pixels, because the X-ray energy (27.2–31.7 keV) of Te is larger than the binding energy (−26.7 keV) of the Cd K shell electron. This means that the Cd K-X-rays are also generated as events progress. In this case, the Cd K-X-rays return to the pixel of interest. A series of events occur instantaneously and it ends within a time shorter than the time resolution of the detector electronics; namely, the electric signal from the pixel of interest cannot distinguish each event, and the pixel puts out a signal equal to the total energy. The total absorbed energy is estimated to be E-K(Te) + K(Cd), where E, K(Te), and K(Cd) represent the energy of incident X-rays, the energy of Te K-X-rays, and energy of Cd K-X-rays, respectively. All of the interactions and corresponding absorbed energies are listed in Table 3.2. In this table, "Escape(Te)" and "Escape(Cd)" represent EPs of the characteristic X-ray escapes. Note that the energies of Cd K-X-rays ($K_{\alpha 1}$, $K_{\alpha 2}$, $K_{\beta 1}$, $K_{\beta 2}$, $K_{\beta 3}$) are superimposed on the EPs of the Te K-X-rays (E-Te($K_{\alpha 1}$), E-Te($K_{\alpha 2}$), E-Te($K_{\beta 1}$, $K_{\beta 3}$), E-Te($K_{\beta 2}$)), and all combinations appear.

The above information reveals that the response function strongly depends on the size of the X-ray irradiation field. In order to estimate the dependency of the irradiation field size, we present the results of a Monte-Carlo simulation [24]. Figure 3.20 shows the simulation conditions and typical results. We selected the parameters for a monolithic CZT detector having a size of 10 mm width, 10 mm length, and 1.5 mm thickness. As shown in the inset, pixel sizes (side length: t) varied from 50, 100, 200, and 500 μm. The side lengths of the irradiation fields changed (side length L is limited up to 4 mm). The typical results for a $t = 200$ μm pixel detector are presented in Fig. 3.20; the vertical axis shows the ratio of the scatter area (blue color in Fig. 3.18) to FEP (red color in Fig. 3.18). We can clearly observe that the small irradiation fields did not achieve equilibrium. Not surprisingly, this result depends on the energy of the incident X-rays. Therefore, dependence was investigated by varying the energy of the incident X-ray from 20 keV to 120 keV. From the results, the irradiation field having a side length of approximately 4 mm was found to be adequate.

Figure 3.21 shows a comparison of response functions for different pixel sizes: the results for 50 μm, 100 μm, 200 μm, and 500 μm are presented. Although there are many peaks in addition to the FEP, they can be easily identified using Table 3.2. The response function for the 50 μm pixel has a relatively small FEP and large scattered areas (EPs and CEs) when compared to those of larger pixels. In recent medical diagnosis, detectors having a pixel size of approximately 200 μm are being used [26], therefore in the following description, we will present the results for a 200 μm pixel size. Figure 3.22 shows a comparison of the response functions for incident energies at 40, 60, 80, and 100 keV. In the graphs, characteristic X-rays for Cd and Te are observed at the energy regions between 20 and 30 keV, while the EPs have variable values depending on the incident energies. Here, we define the response function represented in Fig. 3.22 as follows:

$$\mathbf{R}^{(1)} : \text{response function taken into consideration the X-ray transportations.} \quad (3.7)$$

Table 3.2 Phenomena and corresponding energies for the interactions of escape peaks in a multi-pixel-type CZT detector

Phenomena	Description	Energy [keV]
Escape (Te)	E-Te($K_{\alpha1}$)	E-27.5
	E-Te($K_{\alpha2}$)	E-27.2
	E-Te($K_{\beta1}$, $K_{\beta3}$)	E-31.0
	E-Te($K_{\beta2}$)	E-31.7
Escape (Cd)	E-Cd($K_{\alpha1}$)	E-23.2
	E-Cd($K_{\alpha2}$)	E-23.0
	E-Cd($K_{\beta1}$,$K_{\beta3}$)	E-26.1
	E-Cd(K_{b2})	E-26.6
Escape (Te)+ Characteristic X-rays (Cd)	E-Te($K_{\alpha1}$)+Cd($K_{\alpha1}$)	E-27.5+23.2 = E-4.3
	E-Te($K_{\alpha1}$)+Cd($K_{\alpha2}$)	E-27.5+23.0 = E-4.5
	E-Te($K_{\alpha1}$)+Cd($K_{\beta1}$,$K_{\beta3}$)	E-27.5+26.1 = E-1.4
	E-Te($K_{\alpha1}$)+Cd($K_{\beta2}$)	E-27.5+26.6 = E-0.9
	E-Te($K_{\alpha2}$)+Cd($K_{\alpha1}$)	E-27.2+23.2 = E-4.0
	E-Te($K_{\alpha2}$)+Cd($K_{\alpha2}$)	E-27.2+23.0 = E-4.2
	E-Te($K_{\alpha2}$)+Cd($K_{\beta1}$,$K_{\beta3}$)	E-27.2+26.1 = E-1.1
	E-Te($K_{\alpha2}$)+Cd($K_{\beta2}$)	E-27.2+26.6 = E-0.6
	E-Te($K_{\beta1}$, $K_{\beta3}$)+Cd($K_{\alpha1}$)	E-31.0+23.2 = E-7.8
	E-Te($K_{\beta1}$, $K_{\beta3}$)+Cd($K_{\alpha2}$)	E-31.0+23.0 = E-8.0
	E-Te($K_{\beta1}$, $K_{\beta3}$)+Cd($K_{\beta1}$,$K_{\beta3}$)	E-31.0+26.1 = E-4.9
	E-Te($K_{\beta1}$, $K_{\beta3}$)+Cd($K_{\beta2}$)	E-31.0+26.6 = E-4.4
	E-Te($K_{\beta2}$)+Cd($K_{\alpha1}$)	E-31.7+23.2 = E-8.5
	E-Te($K_{\beta2}$)+Cd($K_{\alpha2}$)	E-31.7+23.0 = E-8.7
	E-Te($K_{\beta2}$)+Cd($K_{\beta1}$,$K_{\beta3}$)	E-31.7+26.1 = E-5.6
	E-Te($K_{\beta2}$)+Cd($K_{\beta2}$)	E-31.7+26.6 = E-5.1
Characteristic X-rays	Te($K_{\alpha1}$), Te($K_{\alpha2}$)	27.5, 27.2
	Te($K_{\beta1}$, $K_{\beta3}$), Te($K_{\beta2}$)	31.0, 31.7
	Cd($K_{\alpha1}$), Cd($K_{\alpha2}$)	23.2, 23.0
	Cd($K_{\beta1}$, $K_{\beta3}$), Cd($K_{\beta2}$)	26.1, 26.6
Escape of X-rays(Cd) from characteristic X-rays(Te)	Te($K_{\alpha1}$)−Cd($K_{\alpha1}$)	27.5−23.2 = 4.3
	Te($K_{\alpha1}$)−Cd($K_{\alpha2}$)	27.5−23.0 = 4.5
	Te($K_{\alpha1}$)−Cd($K_{\beta1}$, $K_{\beta3}$)	27.5−26.1 = 1.4
	Te($K_{\alpha1}$)−Cd($K_{\beta2}$)	27.5−26.6 = 0.9
	Te($K_{\alpha2}$)−Cd($K_{\alpha1}$)	27.2−23.2 = 4.0
	Te($K_{\alpha2}$)−Cd($K_{\alpha2}$)	27.2−23.0 = 4.2
	Te($K_{\alpha2}$)−Cd($K_{\beta1}$, $K_{\beta3}$)	27.2−26.1 = 1.1
	Te($K_{\alpha2}$)−Cd($K_{\beta2}$)	27.2−26.6 = 1.1
	Te($K_{\beta1}$, $K_{\beta3}$)−Cd($K_{\alpha1}$)	31.0−23.2 = 7.8
	Te($K_{\beta1}$, $K_{\beta3}$)−Cd($K_{\alpha2}$)	31.0−23.0 = 8.0
	Te($K_{\beta1}$, $K_{\beta3}$)−Cd($K_{\beta1}$, $K_{\beta3}$)	31.0−26.1 = 4.9
	Te($K_{\beta1}$, $K_{\beta3}$)−Cd($K_{\beta2}$)	31.0−26.6 = 4.4
	Te($K_{\beta2}$)−Cd($K_{\alpha1}$)	31.7−23.2 = 8.5
	Te($K_{\beta2}$)−Cd($K_{\alpha2}$)	31.7−23.0 = 8.7
	Te($K_{\beta2}$)−Cd($K_{\beta1}$, $K_{\beta3}$)	31.7−26.1 = 5.6
	Te($K_{\beta2}$)−Cd($K_{\beta2}$)	31.7−26.6 = 5.1

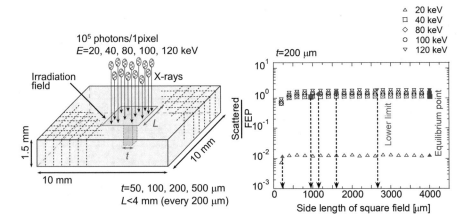

Fig. 3.20 Demonstration of an estimation of the proper-sized irradiation field. In our application, we want to use a common response function which is at equilibrium of secondary produced radiations. Therefore, we should know the proper irradiation field size to estimate response functions

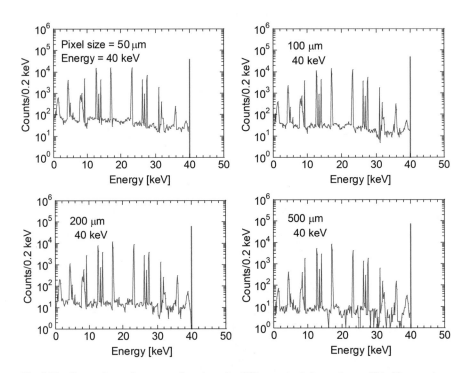

Fig. 3.21 Comparison of response functions for different pixel sizes when a 40 keV monochromatic X-ray is incident to the CZT detector. These spectra were calculated using Monte-Carlo simulation

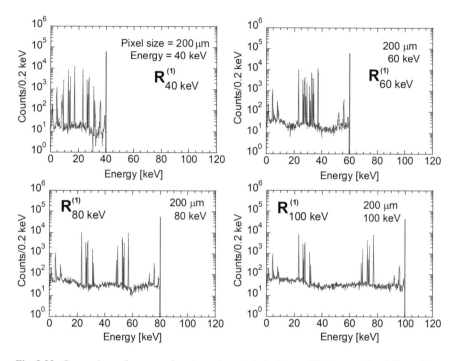

Fig. 3.22 Comparison of response functions of a multi-pixel-type CZT detector for different incident energies. These spectra were calculated using Monte-Carlo simulation

Next, we will discuss the quantitative analysis results for each component of the response function. Figure 3.23 shows a ratio comparison of each component in the response functions for different pixel sizes. The ratio is normalized by the number of incident X-rays for the pixel of interest. Compared with the results for the single-probe-type CdTe detector (see Fig. 3.16), the sum of the ratios for a multi-pixel-type ERPCD increases to over 1.0. These phenomena are caused by contamination from scattered X-rays and characteristic X-rays which are generated in adjacent pixels. The contamination rate rapidly decreases as the pixel size becomes larger. It is generally known that a small pixel size is good for obtaining better special resolution, but the above result indicates the importance of the valance between pixel size and the quality of response function.

Here, we will present several examples of predicted X-ray spectra which are calculated by Monte-Carlo simulation. The original X-ray spectra were reproduced by semi-empirical formulas [25, 27, 28]. These original X-ray spectra are then put into the simulation code, and the response (spectrum: absorbed energy distribution) is calculated. In the calculated spectra, the effect of the response function is clearly observed. Figure 3.24 shows a comparison of predicted X-ray spectra for tube voltages 40, 60, and 80 kV with the original spectra. Black data show original X-ray spectra which are generated from the X-ray tube. Blue and red spectra are folded

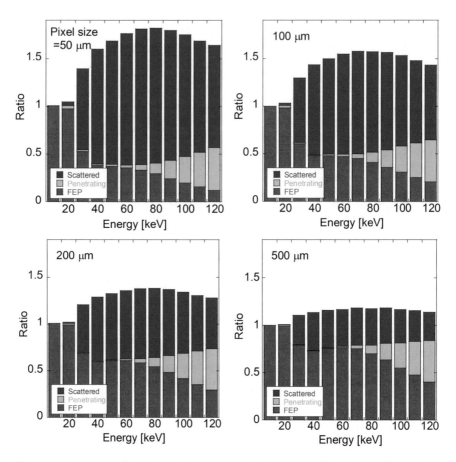

Fig. 3.23 Comparison of the ratio of each component in the response functions for different pixel sizes

spectra for a single-probe-type CdTe detector and a multi-pixel-type CZT detector (pixel size of 200 μm), respectively. Note that these spectra are not similar to the actual spectra, this is because charge collecting processes as well as charge sharing effect are not taken into consideration. The upper, middle, and lower graphs are spectra without an absorber, penetrating a soft-tissue sample (20 mm), and penetrating an aluminum sample (10 mm), respectively. In these spectra, characteristics similar to that of data without an absorber can be seen.

As mentioned in the former sections, the theoretical spectrum (black) and expected spectrum of a single-probe-type CdTe detector (blue) are different, especially around the 27 keV and 32 keV; we remember that these energies correspond to the K-edges of Cd and Te (see Fig. 3.11). Although these noncontiguous behaviors are expected to be seen in the spectra of the single-probe-type CdTe detector, there are no peaks.

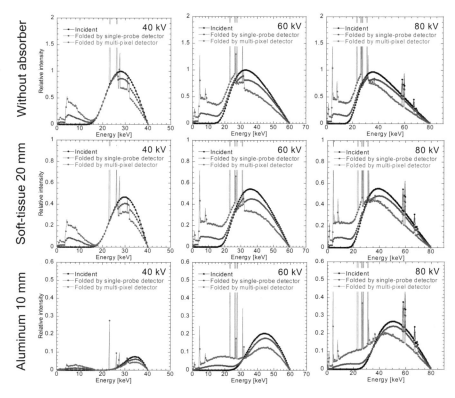

Fig. 3.24 Comparison of predicted X-ray spectra, which are folded with the response functions of a single-probe-type CdTe detector and a multi-pixel-type CZT detector

On the other hand, the multi-pixel-type spectrum (red) reveals specific trends, especially in the 20–30 keV region. Cd and Te characteristic X-rays are clearly observed at energy regions 23 keV and 27 keV, respectively. It is an interesting point that there are many peaks in the predicted X-ray spectra, and we should analyze the cause of these peaks from the consideration of response functions (see Fig. 3.22). We should remember that there are many peaks in the response functions of a multi-pixel-type CZT detector, and the causes of these peaks are classified into two groups when monochromatic X-rays incidence is assumed; one is concerning escape peaks and the other is characteristic X-rays generated in the adjacent pixels. When the polychromatic X-ray spectra (X-ray spectra composed of various "E") are measured, each X-ray generates a different EP. Therefore, the EPs are not be observed in the predicted X-ray spectra. On the other hand, characteristic X-rays are commonly observed regardless of the energy of the incident X-rays. Therefore, the peaks related to characteristic X-rays are strongly enhanced in the X-ray spectra.

3.4.3   Consideration of Charge Sharing Effect and Energy Resolution

In the former section, we explained the response function $\mathbf{R}^{(1)}$ for considering the photon transportation process in the detector, and in addition to this consideration, we should consider the effect of the charge transporting process. Although the former consideration of the response function is based on ideal conditions in which a generated charge cloud is completely absorbed by a pixel, as described in Fig. 3.4, we should look at the charge transport process in a more realistic situation. Analysis based on physics during the charge transporting processes has been described elsewhere. In this book, we do not want to analyze the charge transport process in the detector, but want to use the resulting X-ray spectra measured with an ERPCD for proposing a new analytical procedure. The charge transportation process leads to additional disturbances in the response functions and these are known to be the charge sharing effect [29–31] and energy resolution.

As shown in Fig. 3.25, we define two parts of response functions:

$$1. \ \mathbf{r}_c^{(2)} : \text{peak component and other parts,} \tag{3.8}$$

$$2. \ \mathbf{r}_e^{(2)} : \text{Gaussian function.} \tag{3.9}$$

Here, $\mathbf{r}_c^{(2)}$ and $\mathbf{r}_e^{(2)}$ are response functions to reproduce charge sharing and energy resolution, respectively. $\mathbf{r}_c^{(2)}$ consists of the peak component and the other component which is described as a flat distribution function. "The other" parts are caused by the following physical phenomena. Generated charges cannot be completely collected by the pixel of interest, and the charges are dispersed to the adjacent pixels, but it is possible to assume that various collection rates may occur. We can also imagine that some charge escapes from the pixel of interest and some of this charge enters into the pixel of interest. Considering this, it is expected that "the other" part can be expressed as a symmetrical function, and the functional form is assumed to be greatly affected by the pixel size and incident area of the X-rays against the pixel. It will be very difficult to estimate this function exactly, therefore we decide to use the most simplest symmetric function of a constant. On the other hand, $\mathbf{r}_e^{(2)}$ contains many fluctuating factors such as statistical fluctuation and electrical noise. Since they fluctuate randomly with respect to the center value, we think it is suitable to describe this as a Gaussian function. At this time, the variance in the Gaussian function becomes a parameter that must be determined experimentally. Even if it takes a lot of effort to obtain the response function from an experimental approach, the experiment will be meaningless if the purpose of this experiment is unclear. Since it is not realistic to measure these response functions exactly, we would like to propose a method of optimizing the parameters so as to match the measured X-ray/gamma-ray spectrum. As described later, the optimized parameters used in these response functions are 35% peak efficiencies for the $\mathbf{r}_c^{(2)}$ value and 5%

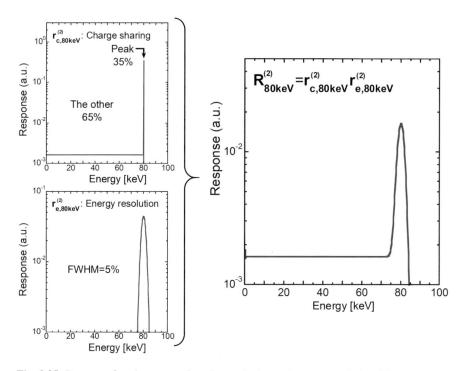

Fig. 3.25 Response function concerning charge sharing and energy resolution. The parameters used in these response functions are optimized so as to reproduce X-ray spectra which are measured with our detector

full width at half maximum (FWHM) for $\mathbf{r}_e^{(2)}$. We also determined energy dependence of energy resolution using a well-known formula; the relative deviation of energy resolution δE is proportional to $1/\sqrt{E}$. The figures on the left in Fig. 3.25 show response functions of $\mathbf{r}_c^{(2)}$ and $\mathbf{r}_e^{(2)}$. We can see the effect of these response functions. The $\mathbf{r}_c^{(2)}$ works to divide the peak component into the other peak components; the peak counts remain at 35% of the peak, and the other 65% is dispersed into the other peak components. The effect of $\mathbf{r}_e^{(2)}$ is dispersal.

The product of these response functions is defined as follows:

$$\mathbf{R}^{(2)} = \mathbf{r}_c^{(2)} \times \mathbf{r}_e^{(2)}, \tag{3.10}$$

where $\mathbf{R}^{(2)}$ is a response function taken into consideration charge sharing and energy resolution. The right figure in Fig. 3.25 shows the $\mathbf{R}^{(2)}$ value of a multi-pixel-type CZT detector using 80 keV monochromatic X-rays.

3.4.4   Reproduction of X-ray Spectra Measured with a Multi-Pixel-Type CZT Detector

Using the response functions $\mathbf{R}^{(1)}$ (Eq. (3.7)) and $\mathbf{R}^{(2)}$ (Eq. (3.10)), we can reproduce X-ray spectra under actual conditions. Figure 3.26 shows the concept of reproducing X-ray spectra using response functions. Case 1 shows an ideal case, in which the total energy of incident X-rays is completely absorbed by one pixel, therefore the obtained X-ray spectrum becomes the same as the incident spectrum. Case 2 shows a semi-realistic situation in which transportations of X-rays with scattering X-rays and generation of characteristic X-rays are involved in the determination of response function $\mathbf{R}^{(1)}$. Case 3 represents an actual case in which both the $\mathbf{R}^{(1)}$ and $\mathbf{R}^{(2)}$ are taken into consideration. As presented in the lower figure, the effect of $\mathbf{R}^{(2)}$ on the obtained spectrum is very large; we can see that the shape of the X-ray spectrum folded by $\mathbf{R}^{(1)}\mathbf{R}^{(2)}$ is much different than that of the ideal X-ray spectrum.

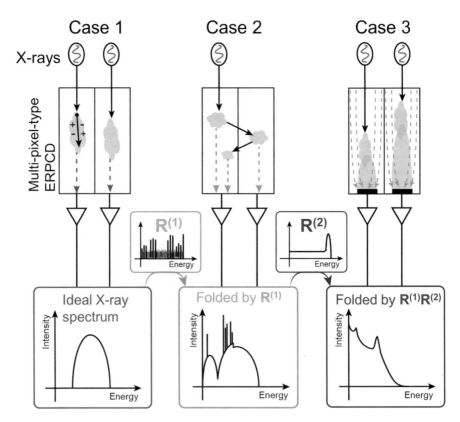

Fig. 3.26 Comparison of three X-ray spectra: case 1, an ideal response in which the energy of incident X-ray is completely absorbed by one pixel; case 2, a response in which transportations of scattering X-rays are taken into consideration; and case 3 an actual response in which charge sharing effect and effect of energy resolution are additionally accounted for

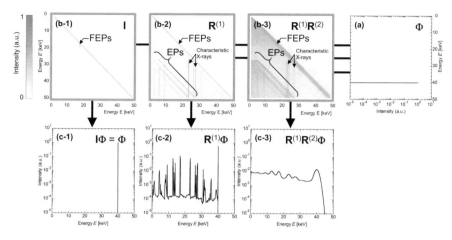

Fig. 3.27 The procedure to calculate the expected spectra including response functions. (**a**) shows an incident X-ray spectrum. (**b-1**), (**b-2**) and (**b-3**) represent response functions. (**c-1**) shows an ideal case, in which the ideal response function **I** is applied. (**c-2**) shows a case in which only response function $\mathbf{R}^{(1)}$ is applied. (**c-3**) represents the actual case, in which response functions $\mathbf{R}^{(1)}$ and $\mathbf{R}^{(2)}$ are applied. To make it easier to understand, we demonstrated a situation using monochromatic X-rays

Here we explain the detailed process of calculating folded spectra while taking into consideration the response functions $\mathbf{R}^{(1)}$ and $\mathbf{R}^{(2)}$. Figure 3.27 shows a schematic drawing to explain the procedure used to reproduce X-ray spectra obtained using a multi-pixel-type ERPCD. First, as shown in Fig. 3.27a, ideal spectrum $\mathbf{\Phi}$ was prepared as follows:

$$\mathbf{\Phi} = \begin{pmatrix} \Phi_1 \\ \Phi_2 \\ \vdots \\ \Phi_j \\ \vdots \end{pmatrix}. \tag{3.11}$$

In this case for easy understanding, we will explain the procedure using $\mathbf{\Phi}$ in which monochromatic X-ray is used. Then, we will present a mathematical description for reproducing the spectra related to cases 1, 2, and 3 in Fig. 3.26. First, the detector responses for cases 1, 2, and 3 are expressed by the following matrices. For case 1, the ideal response of the detector is expressed as **I**.

$$\text{Case 1} : \mathbf{I} = \begin{pmatrix} 1 & 0 & \cdots & \cdots & \cdots & 0 \\ 0 & 1 & \vdots & \vdots & \vdots & \vdots \\ \vdots & \vdots & \ddots & \vdots & \vdots & \vdots \\ \vdots & \vdots & \vdots & \ddots & \vdots & \vdots \\ \vdots & \vdots & \vdots & \vdots & 1 & 0 \\ 0 & \cdots & \cdots & \cdots & 0 & 1 \end{pmatrix}. \tag{3.12}$$

Figure 3.27 (b-1) shows a two-dimensional color map of $\mathbf{I}$. In this plot, it is clearly seen that unit values appear in a diagonal line. Next, we will expand $\mathbf{I}$ to the response function $\mathbf{R}^{(1)}$. $\mathbf{R}^{(1)}$ consists of the response function $\mathbf{R}_{E'}^{(1)}$ of various incident   X-ray energies E'. Then, $\mathbf{R}_{E'}^{(1)}$ is expressed as $\left\{ R_{i,1}^{(1)}, R_{i,2}^{(1)}, \ldots, R_{i,j}^{(1)}, \ldots \right\}$ where i is the element.

$$\text{Case 2} : \mathbf{R}^{(1)} = \begin{pmatrix} R_{1,1}^{(1)} & R_{1,2}^{(1)} & \cdots & \cdots & \cdots \\ R_{2,1}^{(1)} & \ddots & \vdots & \vdots & \vdots & \vdots \\ \vdots & \vdots & \ddots & \vdots & \vdots & \vdots \\ \vdots & \vdots & \vdots & R_{i,j}^{(1)} & \vdots & \vdots \\ \vdots & \vdots & \vdots & \vdots & \ddots & \vdots \\ & \cdots & \cdots & \cdots & \cdots \end{pmatrix}, \tag{3.13}$$

where

$$R_{i,j}^{(1)} = R_{E'(E)}^{(1)}.$$

$\mathbf{R}^{(1)}$ is expressed as a matrix, and we define a row and a column as i and j, respectively; the element $\mathbf{R}^{(1)}$ in the i row and j column is expressed as $R_{i,j}^{(1)}$. $R_{i,j}^{(1)}$ is also expressed by $R_{E'(E)}^{(1)}$ in which E and E' are the energy in the folded spectrum and energy of incident X-rays, respectively. Figure 3.27 (b-2) shows a two-dimensional color map of $\mathbf{R}^{(1)}$. In addition to the strong full energy peak on the diagonal line, characteristic Cd and Te X-rays are clearly observed from 23 to 32 keV. We should note that events corresponding to EPs appear in the energy region below 27 keV, because maximum energy of the X-ray is 50 keV in this case, and the smallest energy of characteristic X-ray is 23 keV (namely, 27 = 50–23 keV). Then, in order to calculate $\mathbf{R}^{(1)}\mathbf{R}^{(2)}$ for determining the response of case 3, $\mathbf{R}^{(2)}$ was defined as follows: matrix $\mathbf{R}^{(2)}$ represents charge sharing and energy resolution. We apply a similar notation rule for $\mathbf{R}^{(1)}$ in Eq. (3.13) to $\mathbf{R}^{(2)}$; namely, matrix $\mathbf{R}^{(2)}$ has elements of $R_{E'}^{(2)}$. In the process of deriving $\mathbf{R}^{(1)}\mathbf{R}^{(2)}$, the element $\mathbf{R}^{(1)}$ in the i row and k column is expressed as $R_{i,k}^{(1)}$, and the element of $\mathbf{R}^{(2)}$ in the k row and j column is expressed as $R_{k,j}^{(2)}$. Then, the response $\mathbf{R}^{(1)}\mathbf{R}^{(2)}$ can be expressed as

Case 3:

$$
\mathbf{R}^{(1)}\mathbf{R}^{(2)} =
\begin{pmatrix}
R_{1,1}^{(1)} & R_{1,2}^{(1)} & \cdots & \cdots & \cdots & \\
R_{2,1}^{(1)} & \ddots & \vdots & \vdots & \vdots & \vdots \\
\vdots & \vdots & \ddots & \vdots & \vdots & \vdots \\
\vdots & \vdots & \vdots & R_{i,k}^{(1)} & \vdots & \vdots \\
\vdots & \vdots & \vdots & \vdots & \ddots & \vdots \\
& \cdots & \cdots & \cdots & \cdots &
\end{pmatrix}
\begin{pmatrix}
R_{1,1}^{(2)} & R_{1,2}^{(2)} & \cdots & \cdots & \cdots & \\
R_{2,1}^{(2)} & \ddots & \vdots & \vdots & \vdots & \vdots \\
\vdots & \vdots & \ddots & \vdots & \vdots & \vdots \\
\vdots & \vdots & \vdots & R_{k,j}^{(2)} & \vdots & \vdots \\
\vdots & \vdots & \vdots & \vdots & \ddots & \vdots \\
& \cdots & \cdots & \cdots & \cdots &
\end{pmatrix}
$$

$$
=
\begin{pmatrix}
R_{1,1}^{(1,2)} & R_{1,2}^{(1,2)} & \cdots & \cdots & \cdots & \\
R_{2,1}^{(1,2)} & \ddots & \vdots & \vdots & \vdots & \vdots \\
\vdots & \vdots & \ddots & \vdots & \vdots & \vdots \\
\vdots & \vdots & \vdots & R_{i,j}^{(1,2)} & \vdots & \vdots \\
\vdots & \vdots & \vdots & \vdots & \ddots & \vdots \\
& \cdots & \cdots & \cdots & \cdots &
\end{pmatrix},
\tag{3.14}
$$

where

$$
R_{i,j}^{(1,2)} = \sum_{k} R_{i,k}^{(1)} R_{k,j}^{(2)}.
$$

In the calculation of $\mathbf{R}^{(1)}\mathbf{R}^{(2)}$, the element $\mathbf{R}^{(1)}\mathbf{R}^{(2)}$ in the i row and j column is expressed as $R_{i,j}^{(1,2)}$. Figure 3.27 (b-3) shows a two-dimensional color map of $\mathbf{R}^{(1)}\mathbf{R}^{(2)}$. The color map is similar to that of $\mathbf{R}^{(1)}$ but the intensities in the low energy region are relatively high. This trend is caused by the charge sharing effect included in $\mathbf{r}_c^{(2)}$. On the other hand, the energy resolution $\mathbf{r}_e^{(2)}$ reduces the sharpness of the observed energy for Cd and Te characteristic X-rays.

In order to obtain folded spectra concerning cases 1, 2, and 3, the following calculations were performed. The spectrum for case 1 can be obtained by

$$
\text{Case 1}: \mathbf{I}\Phi =
\begin{pmatrix}
1 & 0 & \cdots & \cdots & \cdots & 0 \\
0 & 1 & \vdots & \vdots & \vdots & \vdots \\
\vdots & \vdots & \ddots & \vdots & \vdots & \vdots \\
\vdots & \vdots & \vdots & \ddots & \vdots & \vdots \\
\vdots & \vdots & \vdots & \vdots & 1 & 0 \\
0 & \cdots & \cdots & \cdots & 0 & 1
\end{pmatrix}
\begin{pmatrix}
\Phi_1 \\
\Phi_2 \\
\vdots \\
\Phi_j \\
\vdots \\
\end{pmatrix}
=
\begin{pmatrix}
\Phi_1 \\
\Phi_2 \\
\vdots \\
\Phi_j \\
\vdots \\
\end{pmatrix}.
\tag{3.15}
$$

Figure 3.27 (c-1) represents the spectrum for $\mathbf{I}\Phi$. This is the same as the incident spectrum Φ. By replacing $\mathbf{I}$ with $\mathbf{R}^{(1)}$, the following calculation can be performed to obtain the folded spectrum for case 2:

$$\text{Case 2}:\mathbf{R}^{(1)}\mathbf{\Phi}=\begin{pmatrix} R_{1,1}^{(1)} & R_{1,2}^{(1)} & \cdots & \cdots & \cdots \\ R_{2,1}^{(1)} & \ddots & \vdots & \vdots & \vdots & \vdots \\ \vdots & \vdots & \ddots & \vdots & \vdots & \vdots \\ \vdots & \vdots & \vdots & R_{i,j}^{(1)} & \vdots & \vdots \\ \vdots & \vdots & \vdots & \vdots & \ddots & \vdots \\ & \cdots & \cdots & \cdots & \cdots & \end{pmatrix}\begin{pmatrix} \Phi_1 \\ \Phi_2 \\ \vdots \\ \Phi_j \\ \vdots \end{pmatrix}=\begin{pmatrix} \sum_j R_{1,j}^{(1)}\Phi_j \\ \sum_j R_{2,j}^{(1)}\Phi_j \\ \vdots \\ \sum_j R_{i,j}^{(1)}\Phi_j \\ \vdots \end{pmatrix}, \quad (3.16)$$

where

$$\sum_j R_{i,j}^{(1)}\Phi_j = \sum_{E'} R_{E'(E)}^{(1)}\Phi\left(E'\right).$$

$\mathbf{R}^{(1)}\mathbf{\Phi}$ is presented in Fig. 3.27 (c-2). In this spectrum, characteristic Cd and Te X-ray peaks are clearly observed at energies from 23 to 32 keV.

$$\text{Case 3}:\mathbf{R}^{(1)}\mathbf{R}^{(2)}\mathbf{\Phi}=\begin{pmatrix} R_{1,1}^{(1,2)} & R_{1,2}^{(1,2)} & \cdots & \cdots & \cdots \\ R_{2,1}^{(1,2)} & \ddots & \vdots & \vdots & \vdots & \vdots \\ \vdots & \vdots & \ddots & \vdots & \vdots & \vdots \\ \vdots & \vdots & \vdots & R_{i,j}^{(1,2)} & \vdots & \vdots \\ \vdots & \vdots & \vdots & \vdots & \ddots & \vdots \\ & \cdots & \cdots & \cdots & \cdots & \end{pmatrix}\begin{pmatrix} \Phi_1 \\ \Phi_2 \\ \vdots \\ \Phi_j \\ \vdots \end{pmatrix}=\begin{pmatrix} \sum_j R_{1,j}^{(1,2)}\Phi_j \\ \sum_j R_{2,j}^{(1,2)}\Phi_j \\ \vdots \\ \sum_j R_{i,j}^{(1,2)}\Phi_j \\ \vdots \end{pmatrix}, \quad (3.17)$$

where

$$\sum_j R_{i,j}^{(1,2)}\Phi_j = \sum_j \left(\sum_k R_{i,k}^{(1)}R_{k,j}^{(2)}\right)\Phi_j = \sum_{E'}\left(\sum_k R_{i,k}^{(1)}R_{k,E'}^{(2)}\right)\Phi\left(E'\right),$$

where j is replaced by incident energy E'. $\mathbf{R}^{(1)}\mathbf{R}^{(2)}\mathbf{\Phi}$ for incident monochromatic X-ray is shown in Fig. 3.27(c-3). The spectrum $\mathbf{R}^{(1)}\mathbf{R}^{(2)}\mathbf{\Phi}$ has two characteristics: one, there are relatively large intensities in the low energy region, and two, the characteristic Cd and Te X-ray peaks are clearly observed but are spread out due to poor energy resolution. We should note that, after taking into consideration response functions, the X-ray spectrum $\mathbf{R}^{(1)}\mathbf{R}^{(2)}\mathbf{\Phi}$ has a completely different shape when compared with $\mathbf{\Phi}$.

Figure 3.28 shows the predicted results for an ERPCD using CZT when 45 keV, 50 keV, and 55 keV monochromatic X-rays are incident. We would like to understand the results for $\mathbf{R}^{(1)}\mathbf{R}^{(2)}\mathbf{\Phi}$. However, each peak component cannot be clearly discriminated due to the influence of resolution; therefore, we represent the results

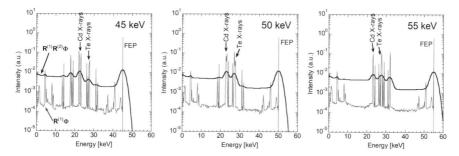

Fig. 3.28 The expected responses when monochromatic X-rays of 45, 50, and 55 keV are incident to the ERPCD

of $\mathbf{R}^{(1)}\mathbf{\Phi}$ as a reference. At first glance, these spectra look very complicated, but they are roughly divided into three components; (1) FEP, (2) EP that changes depending on the energy of incident X-rays, and (3) characteristic Te and Cd X-ray peaks that are unrelated to incident X-ray energy. By comparing the three graphs shown in Fig. 3.28, it can be understood that the characteristic Cd and Te X-ray peaks appear at the same position regardless of incident X-ray energy. Then, because we want to understand the shape of the folded spectrum when continuous X-rays are incident, we will perform a similar calculation by changing the incident X-ray distribution $\mathbf{\Phi}$ from monochromatic X-rays to polychromatic X-rays.

Figure 3.29 is a schematic drawing of the calculation algorithm when the incident X-ray spectrum is changed from monochromatic X-rays to polychromatic X-rays. Figure 3.29(a) is the incident X-ray distribution, which can be calculated using the theoretical formula. Figure 3.29(b-1), (b-2), and (b-3) shows the response functions of $\mathbf{I}$, $\mathbf{R}^{(1)}$, and $\mathbf{R}^{(1)}\mathbf{R}^{(2)}$, respectively, and these response functions are the same as those presented in Fig. 3.27. Figure 3.29(c-1), (c-2), and (c-3) shows folded spectra obtained by the calculation of $\mathbf{I}\mathbf{\Phi}$, $\mathbf{R}^{(1)}\mathbf{\Phi}$, $\mathbf{R}^{(1)}\mathbf{R}^{(2)}\mathbf{\Phi}$, respectively. What we would like to present in this book is that the distribution function $\mathbf{I}\mathbf{\Phi}$, which is treated as an incident X-ray distribution in general books and $\mathbf{R}^{(1)}\mathbf{R}^{(2)}\mathbf{\Phi}$ taking into consideration the actual response of the detector are completely different. In other words, it is not adequate to only provide the novel detector which has energy discriminating ability, but also the potential of the detector should be determined by analyzing the response characteristics of the detector.

Figure 3.30 shows the X-ray spectra at tube voltages of 45 kV, 50 kV, and 55 kV. The solid line is the X-ray spectrum created after considering the response function, and the data shown by open circles are the measured values. They agree very well, and this fact indicates that the calculation method considering the response function is correct. Because our detection system can measure intensities above 20 keV, only data above 20 keV are presented. When folding the X-ray spectra so as to reproduce experimental data, we optimized parameters used for $\mathbf{R}^{(2)}$, and originally calculated results were applied. It is easy to estimate the value of energy resolution because in the spectrum there are intense X-ray peaks related to characteristic Cd and Te X-rays at around 20–28 keV. After optimizing energy resolution, optimization of parameters for charge sharing was performed. Namely, the peak

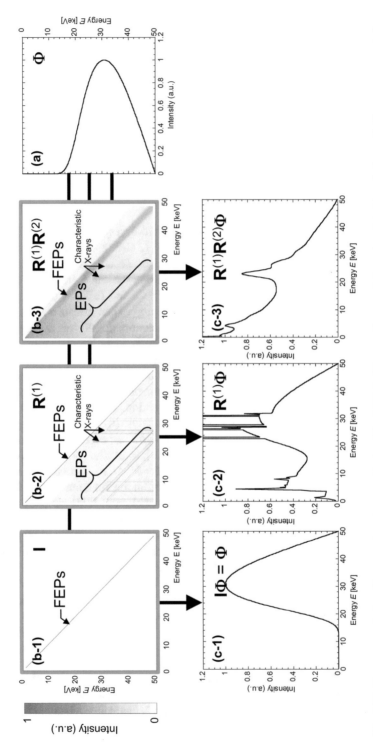

Fig. 3.29 The way to calculate X-ray spectra using a matrix (see Fig. 3.27). In this case, the expected spectra related to polychromatic X-rays are presented

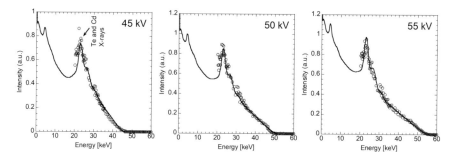

Fig. 3.30 Comparison of X-ray spectra between measured with multi-pixel-type CZT detector and folded with response functions in which X-rays transportations in the detector, charge sharing effect, and effect of energy resolution are taken into consideration. Three different X-ray spectra were measured at 45, 50, and 55 kV with a 2.5 mm aluminum filter

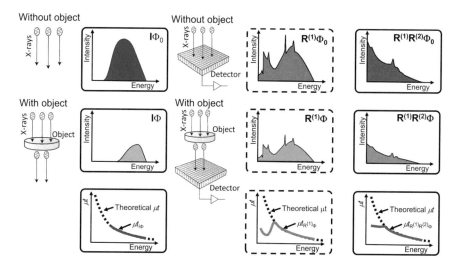

Fig. 3.31 Concept of determining the attenuation factor by calculating the difference of X-ray intensities with and without an object. The effect of response functions is clearly seen in the bottom figures

efficiency (peak parts) parameters were gradually varied, and the value which best represents the experimental value was adopted. At this time, peak efficiency was determined to be 35%. Reproduction of X-ray spectra is important because we can carry out not only various corrections to determine the X-ray attenuation rate but also basic analysis concerning the X-ray spectra.

Here, we will briefly demonstrate how the effect of the response function affects the X-ray spectra by presenting the calculation for attenuation factor. The X-ray spectra shown here are all calculated by simulation, and include the effect of the response function $\mathbf{R}^{(1)}$ described by Eq. (3.7) and $\mathbf{R}^{(2)}$ described by Eq. (3.10). Figure 3.31 shows the analytical concept of attenuation factor μt. By using the X-ray spectra before and after penetrating an object, we can analyze μt from the differences in intensities. As explained in the former section, the relationship between incident X-ray I_0 and penetrating X-ray I is described as $I = I_0 \times \text{Exp}(-\mu t)$. Then, the attenuation factor can be determined using $\mu t = -\ln\left(\dfrac{I}{I_0}\right)$, where μ and t are the linear attenuation coefficient and thickness of an object, respectively. We should note that this formula can be used for the monochromatic X-rays, because μ is a value having energy dependence. From another point of view, the X-ray spectrum can be considered a collection of monochromatic X-rays. Therefore, when we focused attention on the value at a certain energy in the X-ray spectrum, we can apply this formula. Actually, as shown in the left column of Fig. 3.31, the theoretical μt for a wide energy region can be derived from the X-ray spectra of which the effect of detector response can be completely removed.

Next, we want to present a case of using a multi-pixel-type ERPCD. The X-ray spectra measured with a multi-pixel-type CZT detector (folded by $\mathbf{R}^{(1)}$ and/or $\mathbf{R}^{(1)}\mathbf{R}^{(2)}$) can be described with a similar formula:

$$I_D = I_{0,D} \times \exp\left(-\left(\mu t\right)_D\right),$$

$$\left(\mu t\right)_D = -\ln\left(\frac{I_D}{I_{0,D}}\right), \tag{3.18}$$

where $I_{0,D}$, I_D, and $(\mu t)_D$ are observed values which are related to I_0, I, and μt, respectively. As shown in the middle and right columns of Fig. 3.31, the calculation can be performed using two expected spectra; one is without a sample and the other is with a sample.

Figure 3.32 demonstrates the analysis of μt when using an aluminum sample having $\rho t = 1.0$ g/cm^2 as the object of interest. The upper figures show a comparison of predicted X-ray spectra with and without the object. The left, middle, and right columns show the results for the ideal X-ray spectra $\mathbf{I\Phi}$ and folded spectra using $\mathbf{R}^{(1)}$ and $\mathbf{R}^{(1)}\mathbf{R}^{(2)}$, respectively. Data in the second row shows calculated μt values; dashed lines show theoretical values of $\rho t = 0.1$, 1.0 and 10 g/cm^2. As shown on the left, when using the ideal X-ray spectrum $\mathbf{I\Phi}$, we can derive a true μt value. On the other hand, when using the X-ray spectra including the effect of response functions, it is not always possible to calculate correct values over the entire energy range. In the higher energy region, the μt value seems to be in agreement with the theoretical value, but as will be shown later, strictly speaking, they are not in agreement with each other. At this time, we should first roughly focus on a trend seen in the lower energy region; it's clearly seen that the values below 30 keV are not in agreement. This is caused by the effect of the response function. This demonstration indicates that even if the X-ray spectrum is measured, it does not always have information that can achieve quantitative analysis based on the energy of X-rays.

Furthermore, we will analyze the determined μt value from a different viewpoint. Then using additional information, $\rho t = 1.0$ g/cm^2, μ/ρ can be calculated by dividing μt by ρt. In many cases of medical application using plain X-ray, we cannot identify

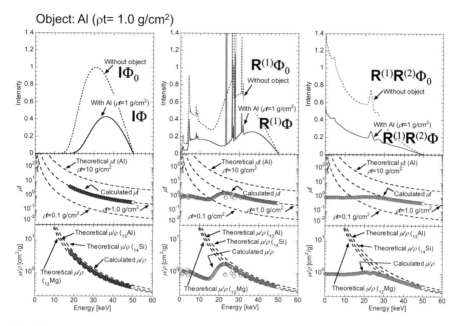

Fig. 3.32 Demonstration of calculating the μt and μ/ρ from the X-ray spectra

mass thickness ρt, therefore the following analysis is just a virtual estimation to explain the importance of considering the response function. In a later section, we will present the idea of deriving an atomic number without thickness "t" information. At this time, we will show one possible method for deriving information of an object. In Fig. 3.32, the determined μ/ρ is shown in the lower figures as a function of energy E. The μt values for different mass thicknesses of 0.1, 1.0, and 10 g/cm^2 can be converted to the same μ/ρ value, therefore we can identify the atomic number of the object regardless of thickness "t" of the object. In the lower figures of Fig. 3.32, theoretical values for μ/ρ for $Z = 12$ (Mg), 13 (Al), and 14 (Si) are also presented. We can see that there is a unique relationship between μ/ρ and the atomic number. From this figure, one can determine the atomic number of an object; namely, when μ/ρ can be determined, the value seems to be converted to atomic number. The lower left figure is the result obtained when ideal X-ray spectra are analyzed. We can identify the true atomic number ($Z = 13$) of the object from the data in the entire energy region. On the other hand, when folded X-ray spectra are used for the analysis, as shown in the middle and right figures, the resulting atomic numbers vary; in the higher energy region, it seems to be near $Z = 13$, but in the lower energy region, it is not. From this demonstration, we can understand that the effect of the response functions of $\mathbf{R}^{(1)}$ and $\mathbf{R}^{(2)}$ on the accuracy of the determined atomic number depends on the energy region. In order to determine the effect of the response functions, we should learn more about response functions and should propose proper analytical procedures to correct the effect.

3.5   Polychromatic X-rays and Effective Energy

Most medical applications using X-ray are based on measuring polychromatic X-rays. This is an important point when considering the analytical procedures used for medical and/or industrial applications. Namely, a novel analysis procedure should be developed that is based on polychromatic X-rays. When using polychromatic X-rays for these analyses, it is difficult to separate various physical phenomena related to various X-rays having different energies. When analyzing a continuous X-ray spectrum as a single energy spectrum, effective and/or averaged energies of X-ray spectra are used. In this section, we will describe the concept of analysis in which continuous X-rays are converted into representative energy.

First of all, the half-value layer (HVL) should be defined. This is achieved by substituting t = HVL and I = $I_0/2$ into the well-known attenuation formula ($I = I_0\mathrm{Exp}(-\mu t)$), and then the following relationship can be derived.

$$\mathrm{HVL} = \frac{\ln(2)}{\mu} \tag{3.19}$$

When we use this equation for the determination of HVL, it should be noted that X-ray intensity (I_0 and I) is measured with an ionization chamber. Using HVL, we can derive the effective energy of polychromatic X-rays. Figure 3.33 shows the relationship between the half-value layer and the energy of a monochromatic X-ray. Here, we use an ideal monochromatic X-ray to understand the relationship between HVL and X-ray energy. When measuring HVL using aluminum, we can calculate the linear attenuation coefficient μ using the above equation. It is important to remember that μ is uniquely determined by the physical property of aluminum. The relationship between μ and energy E is shown in the inset of Fig. 3.33. When summarizing this logical thinking, we can derive a unique relationship between HVL and E. We should be concerned that this relationship can be applied to the measured data using monochromatic X-rays.

Next, we will extend this idea of monochromatic X-rays to the analysis of polychromatic X-rays. Figure 3.34 shows the procedure used to determine the effective energy E_{eff} from the measured HVL when using polychromatic X-rays. Generally, the HVL of polychromatic X-rays is measured with an ionization chamber; the measured value using the ionization chamber is the amount of ionization per unit mass [C/kg], and this value is essentially equivalent to kerma [J/kg]. Then, using the relationship between HVL and energy of monochromatic X-rays, the measured HVL value using polychromatic X-rays can be converted to E_{eff}. In Fig. 3.34, the inset shows X-ray spectra that represent intensity (counts) and kerma. The method presented here is general, but it should be noted that there is an assumption in the methodology; polychromatic X-rays may be treated as a monochromatic X-ray when they are represented by effective energy.

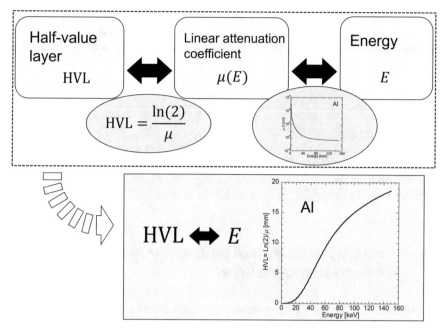

Fig. 3.33 Relationship between the half-value layer (HVL) and energy for monochromatic X-rays

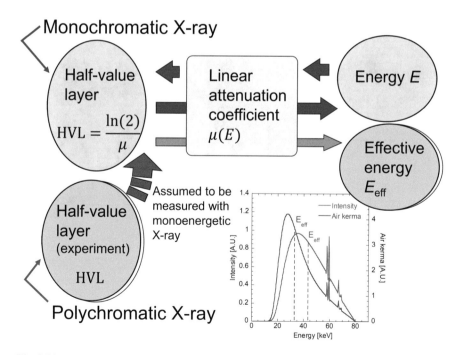

Fig. 3.34 Relationships between the half-value layer and energy of monochromatic and polychromatic X-rays

On the other hand, photon counting techniques use only a part of the X-ray spectrum, therefore the above procedure cannot be applied. In this book, we define the mathematical calculation of the weighted center of the X-ray spectrum as effective energy:

$$E_{eff} = \frac{\int \phi \times E dE}{\int \phi dE}.$$

(3.20)

This equation can be applied, not only to the whole X-ray spectrum, but also to each energy bin in the spectrum. The inset in Fig. 3.34 shows the difference between E_{eff}s determined from the intensity spectrum and kerma spectrum. We can see that the E_{eff} value of the intensity of the spectrum is higher than that of the kerma spectrum.

3.6   Novel Correction Method for Detector Response and Beam Hardening Effect

Based on the former descriptions, we can derive E_{eff} from an X-ray spectrum. In this section, we will consider the variance of E_{eff} when X-rays penetrate objects consisting of different atomic numbers.

Figure 3.35 shows the concept of explaining X-ray penetration rates for different materials. The left figure shows X-ray spectra penetrating air; in this case, attenuation of air is negligible. The corresponding spectrum can be used as a reference. The

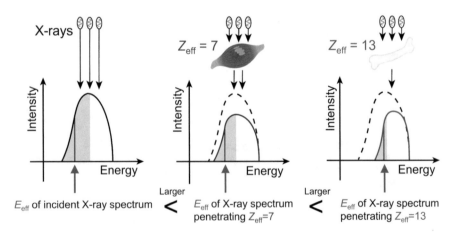

E_{eff}: effective energy in certain energy bin (pink colored area)

Fig. 3.35 Concept of beam hardening effect. The left figure shows X-ray spectrum for reference. Middle and right figures show the X-ray spectra penetrating soft tissue ($Z = 7$) and bone ($Z = 13$), respectively. It can be seen that effective energy (E_{eff}) of the X-ray spectra in a certain energy bin becomes higher

shaded area is set to be a certain energy bin which is located in the lower region in the present case, and the red solid line represents E_{eff} for the energy bin. The middle figure shows the spectrum of X-ray penetrating soft-tissue material (effective atomic number: $Z_{eff} = 7$). It can be seen that the X-ray spectrum in the lower energy region is much more attenuated when compared with the higher region. As shown in right figure, this trend increases when X-rays penetrate bone, which has an Z_{eff} of 13. As presented in Fig. 3.35, the E_{eff} of the penetrating X-ray spectra (middle: $Z_{eff} = 7$, right: $Z_{eff} = 13$) varies with object, and this is called the "beam hardening effect." The analytical procedure we propose in this book is based on the calculation of μ for a certain energy bin. At this time, the correction for the beam hardening effect is important because the effect varies the E_{eff}. When E_{eff} is varied, the corresponding μ is also changed. In other words, when the beam hardening effect can be properly corrected, we can analyze the effective atomic number Z_{eff} of an object using the relationship between E_{eff} and μ.

In an ideal explanation of a photon counting detector, we explained that the energy of incident photons can be analyzed individually, but in an actual case, we can use only a few energy bins because of limitations in engineering technology and consideration must be paid to reducing statistical fluctuation. Namely, when we set an energy bin with an energy width of approximately 10 keV, we should analyze the effective energy of X-rays in this energy bin. Furthermore, when performing an X-ray examination for medical use, we should use polychromatic X-rays which have a wide energy spectrum. In order to perform highly accurate analysis of X-ray attenuation, we propose a correction method for the beam hardening effect in the energy bin of interest of an X-ray spectrum [4–10].

Here, we explain the procedure to correct the beam hardening effect, when the X-ray spectrum was measured with an ERPCD having three energy bins. Interestingly, this beam hardening correction technique can also correct for the response characteristics of the detector simultaneously. Figure 3.36 shows the concept in which correction of the beam hardening effect for certain energy bins is demonstrated. Note that this demonstration is for the beam hardening effect caused by penetrating an aluminum object. This is just a specific case and this concept cannot be applied to all cases. In the upper left figure, we can see the differences between X-ray spectra for $\mathbf{\Phi}$ and $\mathbf{\Phi}_0$. In the upper right figure, we can see the differences in X-ray spectra of $\mathbf{R}^{(1)}\mathbf{R}^{(2)}\mathbf{\Phi}$ and $\mathbf{R}^{(1)}\mathbf{R}^{(2)}\mathbf{\Phi}_0$. As explained above, the latter spectra include two effects; one is the beam hardening effect and the other is detector responses. The attenuation factor for monochromatic X-rays of $\mathbf{\Phi}$ and $\mathbf{\Phi}_0$ can be determined by the following equation.

Monochromatic X-ray using $\mathbf{\Phi}$:

$$\left(\mu t\right)_{cor} = \ln\left(\frac{I_0}{I}\right). \tag{3.21}$$

On the other hand, the actual attenuation factor for polychromatic X-rays of $\mathbf{R}^{(1)}\mathbf{R}^{(2)}\mathbf{\Phi}$ and $\mathbf{R}^{(1)}\mathbf{R}^{(2)}\mathbf{\Phi}_0$ can be derived by the following formula.

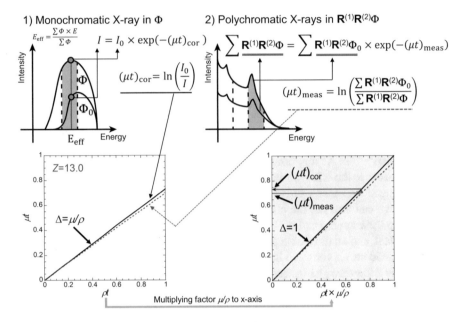

Fig. 3.36 How to perform μt correction. The upper figures show the comparison of monochromatic and polychromatic X-ray spectra note that the monochromatic X-ray spectra do not include response function, but the polychromatic X-ray spectra include response function. The lower figure shows beam hardening correction curves. The black and red lines correspond to monochromatic and polychromatic X-rays, respectively. The arrows in the right figures show the path of correction; namely, the original value (red arrow) is converted to the corrected value (black arrow)

Polychromatic X-rays using $\mathbf{R}^{(1)}\mathbf{R}^{(2)}\mathbf{\Phi}$:

$$\left(\mu t\right)_{\text{meas}} = \ln\left(\frac{\sum \mathbf{R}^{(1)}\mathbf{R}^{(2)}\mathbf{\Phi}_0}{\sum \mathbf{R}^{(1)}\mathbf{R}^{(2)}\mathbf{\Phi}}\right). \tag{3.22}$$

The lower left figure in Fig. 3.36 shows the relationship between mass thickness ρt and attenuation factor μt. When monochromatic X-rays are measured, the relationship of ρt versus $\mu t = (\mu t)_{\text{cor}}$ is a straight line; the gradient of this line becomes μ/ρ, and it is obvious that this value is uniquely determined against certain effective energy regardless of the object's thickness. In order to derive this relationship for monochromatic X-rays, we calculate effective energy in a certain energy bin using Eq. (3.20). Remember that the equation does not include the effect of the response function; in other words, the relationship between ρt and μt can be derived from the ideal incident X-ray spectra and those penetrating an object. This is an important point because this relationship becomes a reference value. Namely, we want to propose the correction method, in which actual attenuations of polychromatic X-rays related to $\mathbf{R}^{(1)}\mathbf{R}^{(2)}\mathbf{\Phi}$ can be converted into ideal attenuations of monochromatic X-rays related to $\mathbf{\Phi}$.

When polychromatic X-rays are used, the effective energy varies with object thickness because lower energy X-rays are easily attenuated by the object. In this case, the relationship of ρt versus μt is a curve. The idea of beam hardening correction is to correct the experimentally obtained μt $((\mu t)_{\text{meas}})$ of polychromatic X-rays and convert them to μt $((\mu t)_{\text{cor}})$ of monochromatic X-rays using a plotted line of ρt versus μt [4–10].

By developing this idea, we proposed to use the relationship between $\rho t \times (\mu/\rho) = \mu t$ and μt in order to perform the correction in a clearer way. The y-axis is μt and it is the same as the value plotted in the procedure described above. On the x-axis, the mass attenuation coefficient μ/ρ, which is the known value of an object, is multiplied by ρt. In the right figure of Fig. 3.36, the relationship between $\rho t \times (\mu/\rho) = \mu t$ and μt is presented. It is clearly seen that the line for monochromatic X-rays becomes a line whose gradient equals 1 (Y = X) as presented in black. The red dashed curve shows the relationships between polychromatic X-rays. Using this information, beam hardening corrections can be performed as shown by the arrows in the figure. Namely, the μt value $((\mu t)_{\text{meas}})$ measured with polychromatic X-rays is converted into the μt value $((\mu t)_{\text{cor}})$ for monochromatic X-rays. We should note that this beam hardening correction procedure can be applied when Z_{eff} of the measured object is known.

The following description utilizes similar data to that of the above explanation, but at this time we present data for different energy bins and different objects having atomic numbers between $Z_{\text{eff}} = 5$–15. Figure 3.37 shows the relationship between ρt and μt to display the beam hardening effect for different atomic number materials.

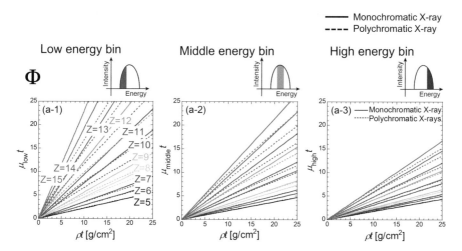

Fig. 3.37 Beam hardening correction curves for different energy bins and materials having different atomic numbers. The solid and broken lines show relationships for monochromatic and polychromatic X-rays, respectively

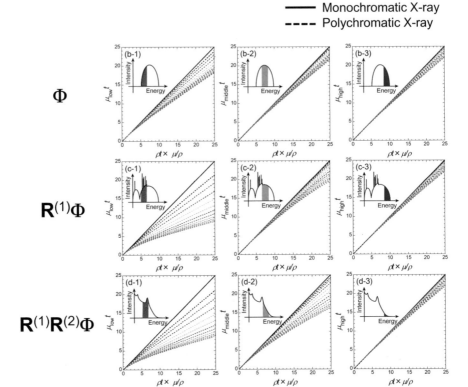

Fig. 3.38 Correction curves for different energy bins related to the different X-ray spectra for Φ, $\mathbf{R}^{(1)}\Phi$, and $\mathbf{R}^{(1)}\mathbf{R}^{(2)}\Phi$. In this figure, the display method which we proposed is presented. In this method, μ/ρ is multiplied by the x-axis

We can identify that the mass attenuation coefficient (μ/ρ) takes a higher value as the atomic number increases. By comparing different energy bins, it is clearly seen that the lower energy bin has higher attenuation factors. Although all data are presented in Fig. 3.37, it is unclear how much changes in beam hardening are due to different atomic numbers. Therefore, we propose expressing the beam hardening effect based on the idea of multiplying with the factor μ/ρ. Figure 3.38 shows beam hardening correction curves in the expressions of $\rho t \times \mu/\rho$ for the x-axis. The upper, middle, and lower graphs show curves when X-ray spectra were assumed to be Φ, $\mathbf{R}^{(1)}\Phi$, and $\mathbf{R}^{(1)}\mathbf{R}^{(2)}\Phi$, respectively. By comparing the results, we found that the amount of correction for the lower energy bin was higher, and this was caused by $\mathbf{R}^{(2)}$. We do not think that the accuracy of the correction for the low energy bin is appropriate because even if the correction can be made, the correction amount is considered to be too large and tended to show statistical fluctuation. In contrast, when focusing attention on middle and high energy bins, the amount of correction is relatively small. This fact shows that the middle and higher energy bins have adequate information to achieve material identification. In our algorithm to derive Z_{eff} of an object, middle and high energy bins are used, and the algorithm is presented in the next section.

References

1. M.J. Willemink, M. Persson, A. Pourmorteza, N.J. Pelc, D. Fleischmann, Photon-counting CT: technical principles and clinical prospects. Radiology **289**, 293–312 (2018). https://doi.org/10.1148/radiol.2018172656
2. X. Wang, D. Meier, K. Taguchi, D.J. Wagenaar, B.E. Patt, E.C. Frey, Material separation in X-ray CT with energy resolved photon-counting detectors. Med. Phys. **38**, 1534–1546 (2011). https://doi.org/10.1118/1.3553401
3. J. Rinkel, G. Beldjoudi, V. Rebuffel, C. Boudou, P. Ouvrier-Buffet, G. Gonon, L. Verger, A. Brambilla, Experimental evaluation of material identification methods with CdTe X-ray spectrometric detector. IEEE Trans. Nucl. Sci. **58**(5), 2371–2377 (2011). https://doi.org/10.1109/TNS.2011.2164266
4. N. Kimoto, H. Hayashi, T. Asahara, E. Tomita, S. Goto, Y. Mihara, Y. Kanazawa, Y. Tamakawa, S. Yamamoto, M. Yamasaki, M. Okada, D. Hashimoto, Novel material identification method using three energy bins of a photon counting detector taking into consideration Z-dependent beam hardening effect correction with the aim of producing an X-ray image with information of effective atomic number, in *Proceedings of 2017 IEEE Nuclear Science Symposium and Medical Imaging Conference (NSS/MIC)*, (2017), p. 18235691. https://doi.org/10.1109/NSSMIC.2017.8533059
5. N. Kimoto, H. Hayashi, T. Asahara, E. Tomita, S. Goto, Y. Kanazawa, S. Yamamoto, M. Okada, M. Yamasaki, Reproduction of response functions of a multi-pixel-type energy-resolved photon counting detector while taking into consideration interaction of X-rays, charge sharing and energy resolution, in *Proceedings of 2018 IEEE Nuclear Science Symposium and Medical Imaging Conference (NSS/MIC)*, (2018). https://doi.org/10.1109/NSSMIC.2018.8824417
6. N. Kimoto, H. Hayashi, T. Asakawa, T. Asahara, T. Maeda, Y. Kanazawa, A. Katsumata, S. Yamamoto, M. Okada, Feasibility study of photon counting detector for producing effective atomic number image, in *Proceedings of 2019 IEEE Nuclear Science Symposium and Medical Imaging Conference (NSS/MIC)*, (2019)
7. T. Asakawa, H. Hayashi, N. Kimoto, T. Asahara, T. Maeda, S. Koyama, S. Yamamoto, M. Okada, Importance of considering the response function of photon counting detectors with the goal of precise material identification, in *Proceedings of 2019 IEEE Nuclear Science Symposium and Medical Imaging Conference (NSS/MIC)*, (2019)
8. M. Sasaki, S. Koyama, Y. Kodera, R. Suzuki, H. Kimura, H. Nishide, M. Mizutani, M. Watanabe, N. Yoshida, H. Hayashi, N. Kimoto, S. Yamamoto, D. Hashimoto, M. Okada, A novel mammographic fusion imaging technique: the first results of tumor tissues detection from resected breast tissues using energy-resolved photon counting detector. SPIE Proc. **10948**, 1094864 (2019). https://doi.org/10.1117/12.2512271
9. N. Kimoto, H. Hayashi, T. Asahara, Y. Mihara, Y. Kanazawa, T. Yamakawa, S. Yamamoto, M. Yamasaki, M. Okada, Precise material identification method based on a photon counting technique with correction of the beam hardening effect in X-ray spectra. Appl. Radiat. Isot. **124**, 16–26 (2017). https://doi.org/10.1016/j.apradiso.2017.01.049
10. N. Kimoto, H. Hayashi, T. Asahara, Y. Kanazawa, T. Yamakawa, S. Yamamoto, M. Yamasaki, M. Okada, Development of a novel method based on a photon counting technique with the aim of precise material identification in clinical X-ray diagnosis, in *Proceedings of SPIE 10132*, (2017), p. 1013239-1-11. https://doi.org/10.1117/12.2253564
11. G.F. Knoll, *Radiation detection and measurement* (John Wiley & Sons, Inc., Hoboken, ISBN-10: 0470649720, 2012)

12. N. Tsoulfanidis, S. Landsberger, *Measurement and detection of radiation* (CRC Press, Boca Raton, ISBN-10: 9781482215496, 2015)
13. K. Maeda, M. Matsumoto, A. Taniguchi, Compton-scattering measurement of diagnostic x-ray spectrum using high-resolution Schottky CdTe detector. Med. Phys. **32**, 1542–1547 (2005). https://doi.org/10.1118/1.1921647
14. S. Miyajima, K. Imagawa, M. Matsumoto, CdZnTe detector in diagnostic X-ray spectroscopy. Med. Phys. **29**, 1421–1429 (2002). https://doi.org/10.1118/1.1485975
15. Y. Kojima, M. Shibata, H. Uno, K. Kawade, A. Taniguchi, Y. Kawase, K. Shizuma, A precise method of $Q\beta$ determination with small HPGe detector in an energy range of 1-9 MeV. Nucl. Inst. Methods Phys. Res. A **458**, 656–669 (2001). https://doi.org/10.1016/S0168-9002(00)00899-8
16. R. Ballabriga, M. Cambell, E. Heijne, X. Llopart, L. Tlustos, W. Wong, Medipix3: A 64 k pixel detector readout chip working in single photon counting mode with improved spectrometric performance. Nucl. Inst. Methods Phys. Res. A **633**, S15–S18 (2011). https://doi.org/10.1016/j.nima.2010.06.108
17. C. Ullberg, M. Urech, N. Weber, A. Engman, A. Redz, F. Henckel, Measurements of a dual-energy fast photon counting detector with integrated charge sharing correction, in *Proceedings of SPIE 8668, 86680P-1-8*, (2013). https://doi.org/10.1117/12.2007892
18. A. Brambilla, P. Ouvrier-Buffet, J. Rinkel, G. Gonon, C. Boudou, CdTe linear pixel x-ray detector with enhanced spectrometric performance for high flux x-ray imaging. IEEE Nucl Sci Conf Record **R18-5**, 4825–4828 (2011)
19. A. Tomal, D.M. Cunha, M. Antoniassi, M.E. Poletti, Response functions of Si(Li), SDD and CdTe detectors for mammographic X-ray spectroscopy. Appl. Radiat. Isot. **70**, 1355–1359 (2012). https://doi.org/10.1016/j.apradiso.2011.11.044
20. E. Storm, H.I. Israel, Photon cross sections from 1 keV to 100 MeV for elements Z = 1 to Z = 100. Nuclear Data Tables **A7**, 565–681 (1970)
21. L. Tlustos, M. Campbell, C. Fröjdh, P. Kostamo, S. Nenonen, Characterisation of an epitaxial GaAs/Medipix2 detector using fluorescence photons. Nucl. Inst. Methods Phys. Res. A **591**, 42–45 (2008). https://doi.org/10.1016/j.nima.2008.03.020
22. T.E. Everhart, P.H. Hoff, Determination of kilovolt electron energy dissipation vs penetration distance in solid materials. J. Appl. Phys. **42**, 5837 (1971). https://doi.org/10.1063/1.1660019
23. R.B. Firestone, V.S. Shirley, *Table of isotopes*, 8th edn. (John Wiley and Sons, Inc., Hoboken, ISBN 0471-14918-7, 1998)
24. H. Hirayama, Y. Namito, Z.F. Bielajew, S.J. Wilderman, W.R. Nelson, The EGS5 code system. KEK Rep **2005-8**, 1–418 (2005)
25. R. Birch, M. Marshall, Computation of bremsstrahlung X-ray spectra and comparison with spectra measured with a Ge(Li) detector. Phys. Med. Biol. **24**(3), 505–517 (1979)
26. R.R. Carlton, A.M. Adler, *Principles of Radiographic Imaging*, 5th edn. (Delmar Cengage Learning, New York, ISBN-13: 978-1-4390-5872-5, 2003)
27. D.M. Tucker, G.T. Barnes, D.P. Chakraborty, Semiempirical model for generating tungsten target X-ray spectra. Med. Phys. **18**(2), 211–218 (1991)
28. J.P. Bissonnette, L.J. Schreiner, A comparison of semiempirical models for generating tungsten target x-ray spectra. Med. Phys. **19**, 579–582 (1992). https://doi.org/10.1118/1.596848
29. C. Ullberg, M. Urech, N. Weber, A. Engman, A. Redz, F. Henckel, Measurements of a dual-energy fast photon counting CdTe detector with integrated charge sharing correction. Proc. SPIE **8668**, 86680P (2013). https://doi.org/10.1117/12.2007892
30. K. Mathieson, M.S. Passmore, P. Seller, M.L. Prydderch, V. O'Shea, R.L. Bates, K.M. Smith, M. Rahman, Charge sharing in silicon pixel detectors. Nucl. Inst. Methods Phys. Res. A **487**, 113–122 (2002). https://doi.org/10.1016/S0168-9002(02)00954-3
31. P. Otfinowski, Spatial resolution and detection efficiency of algorithms for charge sharing compensation in single photon counting hybrid pixel detectors. Nucl. Inst. Methods Phys. Res. A **882**, 91–95 (2017). https://doi.org/10.1016/j.nima.2017.10.092

Chapter 4
Material Identification Method with the Aim of Medical Imaging

4.1 Information Which Can Be Derived from Attenuation Factor

In this section, we would like to consider how atomic number information can be obtained from the information obtained from the X-ray equipment. In the example shown in Fig. 4.1a, it is a concept of a general radiography system, and Fig. 4.1b is that of a computed tomography (CT) system. In general radiography, the amount of X-ray attenuation is determined by μ, which depends on the atom, and the thickness t of the object. The measurable physical quantities are the incident X-ray intensity I_0 and the penetrating X-ray intensity I, and they are in the relation of $\mu t = \ln(I_0/I)$. The μt trend for each element shown in Fig. 4.1a is the value when t is a constant ($t = 1$ cm). Because this physical quantity of μt is strongly affected by the density of the object, they are roughly divided into gases and solids. Looking at it in more detail, we can see that solid materials also have various densities. Density is not a physical quantity that strongly depends only on the atomic number. As a result, information about the atomic number Z cannot be clearly extracted from the measured quantity of μt. On the other hand, for CT examinations, because the detector can be rotated 360° to obtain information related to different projection angles, a two-dimensional μ map can be obtained as an X-ray image. This analysis includes the process of image reconstruction, and the actual CT value is defined as a relative value related to the μ of water. Because CT has a wide range of medical applications and is an essential modality in modern medicine, a great deal of research has been done on this topic. However, we should note that this relative μ value does not completely reflect material information. As shown in Fig. 4.1b, when μ values for different atomic numbers are plotted against energy, the effect of the density of the material has a strong influence; therefore, it is not possible to extract atomic number information contained in μ.

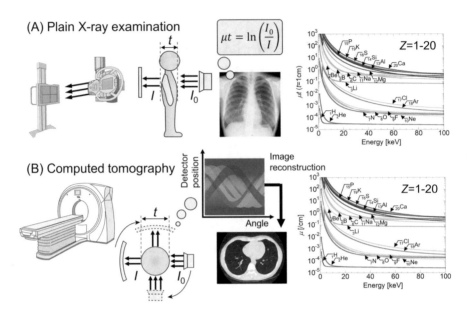

Fig. 4.1 The difference of information obtained for plain X-ray examination and computed tomography

Next, let us consider how atomic number information can be extracted from μt. One of the solutions is to measure μt at multiple energies. In order to show that this solution is a good scientific procedure, we will discuss how μt is related to atomic number. The fact that μt is not a physical quantity determined uniquely by atomic number has already been explained in Fig. 4.1, but here we will explain it again from by focusing on the difference between a single atom and a collection of atoms.

We will start the explanation by assuming that μt was measured by X-rays having a single energy, as shown in Fig. 4.2a. If additional information concerning the thickness "t" of the object can be obtained from other measurements, the following relational expression can be obtained by dividing μt by t:

$$\mu = \frac{\mu t}{t} = \frac{\rho}{A} N_A \times \sigma, \qquad (4.1)$$

where ρ, A, N_A, and σ are density, atomic weight, Avogadro number (6.02×10^{23}), and cross-section, respectively. σ is a value defined for each atom and is the probability of interaction frequency, such as photoelectric effect and Compton scattering effect, etc. Since $\rho N_A / A$ is the number of atoms per 1 cm^3, Eq. (4.1) gives the integral amount, which is the sum of the interactions of all atoms contained within 1 cm^3. Focusing on the atomic number dependence on the parameters included in this equation, the problem is that ρ and A are not smoothly continuous functions which can be uniquely determined concerning the atomic number Z. Furthermore,

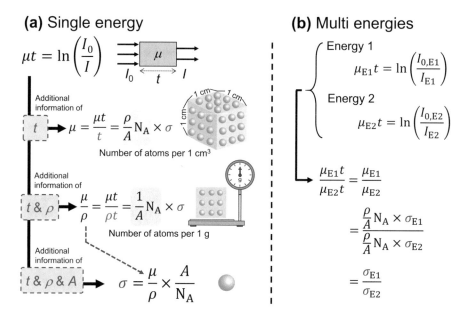

Fig. 4.2 The concept of physical values derived from μt values. Cross-section of σ and how it relates to the atomic number Z of the object. (**a**) When a single energy was used, it is difficult to derive the Z value because μs cannot be separated from μt value without any assumptions. (**b**) On the other hand, when multi-energies are used, the ratio of μs can be determined; this value is strongly related with Z

let us consider another case where information for density ρ is obtained in addition to thickness t. In this case, the following calculation can be performed:

$$\frac{\mu}{\rho} = \frac{\mu t}{\rho t} = \frac{1}{A} N_A \times \sigma. \tag{4.2}$$

The physical quantity being available from a general database is μ/ρ [1], which is called the mass attenuation coefficient. We can calculate μ when multiplying μ/ρ by ρ. Although μ/ρ does not include information about the density of the substance, μ/ρ can be flexibly applied to different density states; therefore, this value is easier to use when compared with μ, which is strongly related to the density of the substance in a standard condition. Because N_A/A represents the number of atoms per unit mass (1 g), it is indicated that the physical quantity obtained from Eq. (4.2) becomes the total cross-section value when the atoms corresponding to 1 g are collected. The physical quantity obtained from Eq. (4.2) does not always depend strongly on atomic number. This consideration may be surprising. In order to understand this fact, it is necessary to understand that atomic weight "A" is not a physical quantity that can be continuously determined with a function of atomic number.

Let us develop the understanding further. If information concerning thickness t, density ρ, and atomic weight A is obtained in addition to the experimental value μt, we can calculate σ by multiplying (μ/ρ) obtained from Eq. (4.2) by using A/N_A as follows:

$$\sigma = \frac{\mu}{\rho} \times \frac{A}{N_A}. \tag{4.3}$$

Since this σ is the cross-section per one atom, it is a unique value for the atomic number Z. Obtaining this value requires information for thickness t, density ρ, and atomic weight A, but this procedure is impractical.

Therefore, as shown in Fig. 4.2b, an analytical procedure that uses the μts of two different energies was developed. This method can be applied to photon counting and dual-energy type imaging methods. Namely, because the attenuation factor μt can be calculated for each energy, the following two physical quantities can be experimentally obtained:

$$\text{Energy1}: \mu_{E1}t = \ln\left(\frac{I_{0,E1}}{I_{E1}}\right), \tag{4.4a}$$

$$\text{Energy2}: \mu_{E2}t = \ln\left(\frac{I_{0,E2}}{I_{E2}}\right). \tag{4.4b}$$

Here, we calculate the ratio of these values, and using Eq. (4.1) we also examine the physical parameter elements included in the calculation for ratio:

$$\frac{\mu_{E1}t}{\mu_{E2}t} = \frac{\mu_{E1}}{\mu_{E2}} = \frac{\frac{\rho}{A}N_A \times \sigma_{E1}}{\frac{\rho}{A}N_A \times \sigma_{E2}} = \frac{\sigma_{E1}}{\sigma_{E2}}, \tag{4.5}$$

where σ_{E1} and σ_{E2} represent cross-sections related to two different energies E_1 and E_2, respectively. When we tried to obtain a relational expression for σ with a single energy X-ray, additional information for t, ρ, and A was necessary. However, by measuring μt using two different energies as shown in Eq. (4.5), it is clearly seen that we can obtain a relational expression containing the only σs, which are strongly related to the atomic number.

Figure 4.3 shows plots of μ, μ/ρ, and σ for elements in the range of $Z = 1-20$. It is implied that the value of interest can be described by a function of atomic number Z if these values for each element change continuously and smoothly. Obviously, μ shown in Fig. 4.3a cannot be described as a function of Z. This is because that ρ and A are included as shown in Eq. (4.1), and they differ greatly for each element. At first glance, μ/ρ shown in Fig. 4.3b seems to be a physical quantity that is uniquely determined by atomic number Z. However, as shown in the enlarged view, it can be

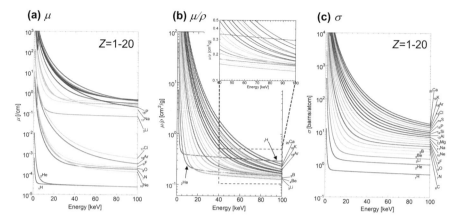

Fig. 4.3 Comparison between μ, μ/ρ, and σ as a function of X-ray energy. The unique relationship between Z and energy can be observed for σ, although clear correlations cannot be seen in the μ and μ/ρ

seen that the intervals are not constant for each element, and the μ/ρ curves for different elements intersect. This fact means that even if μ/ρ is determined for a certain energy, there are cases where an element cannot be identified. The theoretical explanation is that the atomic weight A is not a smooth function with respect to Z, as already explained in Eq. (4.2). On the other hand, as shown in Fig. 4.3c, σs of all atomic numbers are neatly plotted. This fact means that an element of interest can be uniquely identified if σ can be measured at a certain energy. In addition, this relational expression means that it is possible to analyze the effective atomic number of substances consisting of complex materials. It should be noted that although σ contains information related to atomic number Z, σ cannot be analyzed from X-ray images using our method.

Figure 4.4 shows plots of μ, μ/ρ, and σ obtained from each energy bin for the effective energy. Note that this is just a demonstration and μ, μ/ρ, and σ may not always be obtained using an actual ERPCD. Figure 4.4a is the theoretically predicted X-ray spectrum, in which the tube voltage of 50 kV is adopted. The effective energies of the low, middle, and high energy bins are 26.8 keV, 35.7 keV, and 43.3 keV, respectively. Figure 4.4b shows the relationship between μ and Z for three effective energies. As clearly seen in this figure, μ has very complicated behavior with respect to Z. Figure 4.4c demonstrates the dependence of μ/ρ on Z. For atoms lower than $_4$Be, the trend is different and the relationship is not smooth in other places. On the other hand, as shown in Fig. 4.4d, σ becomes a smooth and continuous function of Z. Here, let us explain in a little more detail why μ/ρ does not have a smooth relationship with Z. The left figure in Fig. 4.5 is a nuclear chart, in which X and Y axes show neutron and proton numbers, respectively. The nuclei shown in blue are stable nuclei, and the value shown in the same box indicates abundance. We want to predict the abundance of nuclei using a simple function of Z. At first glance, it seems to follow very complicated rules. The red broken line represents Y = X,

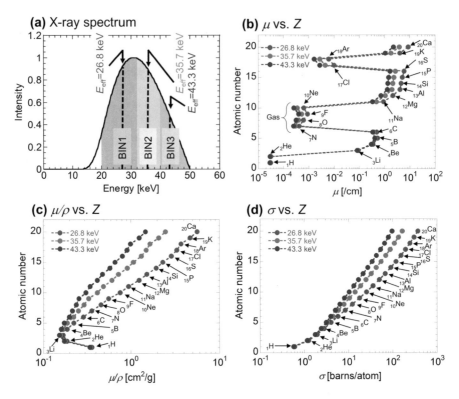

Fig. 4.4 Comparison of the relationships between physical values and atomic number Z. (**a**) shows X-ray spectrum in which three energy bins are used. (**b**) represents μ vs. Z; the μ values are strongly related to the statement of the material. (**c**) shows μ/ρ vs. Z; relatively smooth curves can be seen, but this curve reflects the specific characteristics of each element. (**d**) shows σ vs. Z; a smooth relationship between σ and Z can be obtained

which means that the nuclei located on this line have the same number of protons and neutrons. Stable nuclei of ^{4}He, ^{12}C, ^{14}N, and ^{16}O have the same number of protons and neutrons, and their abundance is relatively large for these isotopes. These nuclei are characterized by having even-numbered nuclei. On the other hand, ^{19}F and ^{23}Na are examples of nuclei that are stable with odd-numbered nuclei. Depending on the atomic number of interest, even-numbered or odd-numbered nuclei become stable, therefore, as a result, atomic weight "A" is not a smooth function of atomic number Z. In particular, this feature is strongly observed in lighter nuclides. The right figure in Fig. 4.5 shows the correlation between atomic weight A and atomic number Z. The existence of various stable nuclei in a relation of $A = 2 \times Z$ and those not being in this line is the reason why A cannot be expressed as a smooth function of Z.

One goal in the development of an ERPCD is to establish a method that can produce a novel X-ray image for material identification. The basic idea for identifying foreign material in normal tissue by using the calculation for attenuation of X-rays has been proposed elsewhere; material identification can be achieved through analy-

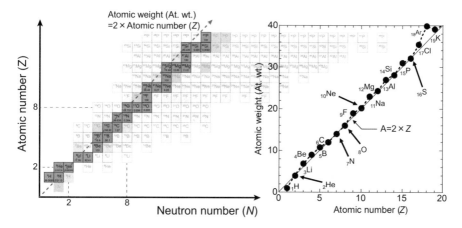

Fig. 4.5 Relationship between atomic weight and atomic number. Because stable nuclei do not always exist on the line described by $A = 2 \times Z$, it is difficult to describe the atomic weight precisely with a function of atomic number Z

sis using linear attenuation coefficients for different atomic numbers [2]. Actually, a good way to analyze material is using the value for "μs." Here, we define μs for different energies as μ_1 and μ_2; using these values for material identification, a two-dimensional scatter plot for μ_1 and μ_2, or ratio of μ_1 and μ_2 (μ_1/μ_2) is usually used. For medical applications, computed tomography (CT) scanners can create a three-dimensional μ map, therefore many researchers consider applying this analytical procedure to perform more accurate analysis of CT data. However, current clinical CT equipment consists of an energy integrating detector (EID) and it will take great effort to replace this detector with an ERPCD. As a first step for clinical studies, dual-energy examinations have been performed [3–5]. Dual-energy CT showed significant results and it was reported that the system has the ability to separate contrast medium (iodine) and calcification quantitatively. This is significant because these materials gave similar imaging densities when using a single energy CT. The use of an energy-resolving photon counting detector for clinical CT is recognized as next-generation type equipment [6]. Although dual-energy CT has been widely applied to clinical equipment, there are some problems which cause a decrease in accuracy. Dual-energy CT uses different X-ray spectra with different tube voltages, and in this concept, each X-ray spectrum has a different effective energy. Because the EID cannot identify effective energies, it is necessary to predict the response of an EID based on the object, and by assuming general clinical situations. In previous studies concerning the development of an EID, some ideas on how to solve the problem have been proposed in detail (Chap. 2 [15]).

4.2   Method to Derive Z_{eff} Value Using ERPCD

Now let us go back to the topic of analysis of attenuation factors measured with an ERPCD. Using the μts for different energy bins and taking into consideration the beam hardening effect, effective atomic numbers can be derived. Figure 4.6 shows the concept of how to derive an effective atomic number (Z_{eff}). As shown in the figure, the algorithm needs the following four processes:

1. calculation of μt,
2. correction of calculated μt under the assumption of Z_{tent},
3. conversion of corrected μt to Z_{eff} under the assumption of Z_{tent},
4. determination of Z_{eff}.

First of all, we determined μt values experimentally. Here, the values are defined as

(a) low energy bin: $(\mu_{low}t)_{meas}$,
(b) middle energy bin: $(\mu_{middle}t)_{meas}$,
(c) high energy bin: $(\mu_{high}t)_{meas}$,

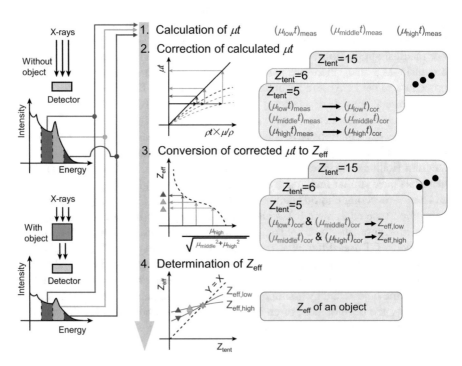

Fig. 4.6 A schematic drawing of an analytic scheme to derive Z_{eff} of an object. From the measured values of μt for different energy bins, differences of μt are properly analyzed to derive the Z_{eff} of an object

where μ_{low}, μ_{middle}, and μ_{high} are linear attenuation coefficients for low, middle, and high energy bins, respectively, and "t" is the thickness of an object. When the correction of beam hardening effect and detector response is performed, the μt values, which are related to polychromatic X-rays, are converted to a value corresponding to monochromatic X-ray as follows:

(a) low energy bin: $(\mu_{\text{low}}t)_{\text{meas}} \rightarrow (\mu_{\text{low}}t)_{\text{cor}}$,
(b) middle energy bin: $(\mu_{\text{middle}}t)_{\text{meas}} \rightarrow (\mu_{\text{middle}}t)_{\text{cor}}$,
(c) high energy bin: $(\mu_{\text{high}}t)_{\text{meas}} \rightarrow (\mu_{\text{high}}t)_{\text{cor}}$,

where subscripts of "meas" and "cor" correspond to measured and corrected values, respectively. The correction needs to be carried out under the assumption of a certain atomic number; in our algorithm described in Fig. 4.6, Z_{eff} was tentatively determined as "Z_{tent}" when the correction is applied. Here, we can easily imagine that the Z_{tent}s of 7.0 and 13.0 are proper settings for beam hardening corrections of soft tissue and bone, respectively. On the other hand, since complex substances containing various atomic numbers are measured in actual medical diagnosis, we assumed Z_{tent}s between 5 and 15 in our algorithm. Then, in order to delete the information for "t" from the μt value, the following calculation is performed:

$$\mu_{\text{low}}^{\dagger} = \frac{(\mu_{\text{low}}t)_{\text{cor}}}{\sqrt{(\mu_{\text{low}}t)_{\text{cor}}^2 + (\mu_{\text{middle}}t)_{\text{cor}}^2}} = \frac{(\mu_{\text{low}})_{\text{cor}}}{\sqrt{(\mu_{\text{low}})_{\text{cor}}^2 + (\mu_{\text{middle}})_{\text{cor}}^2}}. \quad (4.6a)$$

$$\mu_{\text{high}}^{\dagger} = \frac{(\mu_{\text{high}}t)_{\text{cor}}}{\sqrt{(\mu_{\text{middle}}t)_{\text{cor}}^2 + (\mu_{\text{high}}t)_{\text{cor}}^2}} = \frac{(\mu_{\text{high}})_{\text{cor}}}{\sqrt{(\mu_{\text{middle}})_{\text{cor}}^2 + (\mu_{\text{high}})_{\text{cor}}^2}}. \quad (4.6b)$$

In this book, the values $\mu_{\text{low}}^{\dagger}$ and $\mu_{\text{high}}^{\dagger}$ are named "normalized linear attenuation coefficient." Figure 4.7 shows the trend of the normalized linear attenuation coefficient plotted over a range of all atomic numbers. When the atomic number is larger than 20, it means that various Zs can be solutions calculated from a single $\mu^{\dagger}$ and as a result, the proper atomic number cannot be uniquely identified. However, it can be seen that it has sufficient discrimination ability for low atomic number materials which are used in medicine. The problem that needs to be solved is that Z_{tent} used for the correction process and the true value of Z_{eff} for the final result must be determined simultaneously and consistently. The key to solving this problem is the methodology used to estimate Z_{eff} from $\mu_{\text{high}}^{\dagger}$. We will explain why $\mu_{\text{high}}^{\dagger}$ is suitable in the next paragraph.

Using a theoretically calculated database of μ for different atomic number materials (Chap. 3 [20]), datasets $\{\mu_{\text{low}}^{\dagger}, Z_{\text{eff}}\}$ and $\{\mu_{\text{high}}^{\dagger}, Z_{\text{eff}}\}$ which are used as reference curves can be determined. Using an aluminum sample as an example, Fig. 4.8 illustrates the relationship between the energy bin of the X-ray spectrum used for data analysis, μt correction curves, and conversion procedure from normalized linear attenuation factor to Z_{eff}. The data in Fig. 4.8a shows an analytical process when

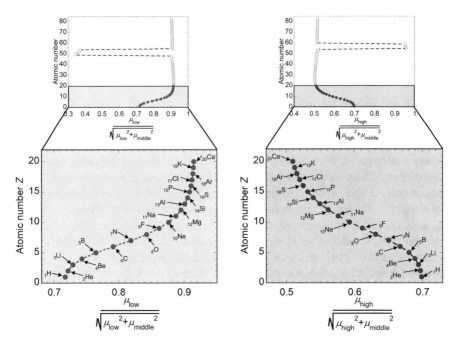

Fig. 4.7 Relationship between normalized linear attenuation factor and atomic number. Using a unique relationship found in the lower Z region, we can identify the effective atomic number of the object using polychromatic X-rays

using the theoretical X-ray spectrum $\mathbf{I\Phi}$. This example shows an ideal condition in which the effect of the detector response function can be ignored. The effect of beam hardening correction on results is actually very small. As it can be seen from the conversion curve for $\mu^\dagger$ to Z_{eff} shown at the bottom, there is not very much difference between the Z_{eff}s when reflecting on the original count and the corrected count. On the other hand, in the example shown in Fig. 4.8b, it is clearly shown that an appropriate correction can be performed by reproducing the distribution of the actually measured X-ray spectra $\mathbf{R}^{(1)}\mathbf{R}^{(2)}\mathbf{\Phi}$ when obtaining $(\mu t)_{\text{cor}}$. Namely, the Z_{eff} calculated using the original count is completely different when compared to the true value of 13, but the true value can be calculated by using $\mu_{\text{low}}{}^\dagger$ and $\mu_{\text{high}}{}^\dagger$ in which corrections for beam hardening and detector response are properly performed. From this example, it can also be understood that the $\mu_{\text{low}}{}^\dagger$ correction value is much greater than that of $\mu_{\text{high}}{}^\dagger$. This procedure indicates that Z_{eff} can be derived when $\mu_{\text{high}}{}^\dagger$ is determined experimentally; however, we should note that the analysis is based on the assumption of using a known Z_{tent} with beam hardening and detector response corrections.

Next, we will explain a procedure to estimate Z_{eff} when measuring an unknown object. In the example, we will measure aluminum ($Z_{\text{eff}} = 13$), and we show that our procedure can derive $Z_{\text{eff}} = 13$ without using the preset value of $Z_{\text{eff}} = 13$. Here, note that $Z_{\text{eff,low}}$ is defined as the value obtained from $\mu_{\text{low}}{}^\dagger$ and it is not the final Z_{eff} value.

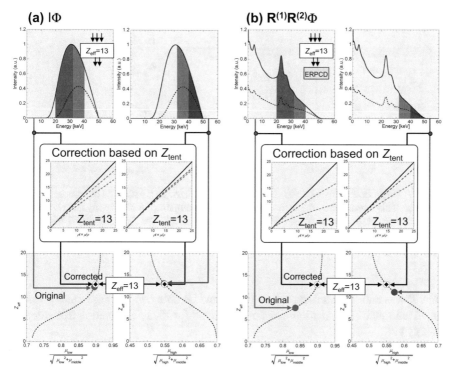

Fig. 4.8 Analytical procedure to derive an effective atomic number. The left and right represent cases of using the theoretical X-ray spectra and expected X-ray spectra, respectively. Because of different shapes in X-ray spectra, correction curves are much different

Similarly, $Z_{eff,high}$ is the Z_{eff} obtained from $\mu_{high}{}^{\dagger}$. As shown in Fig. 4.9a, true correction can be applied to the aluminum sample when we can assume $Z_{tent} = 13$, and then $Z_{eff} = 13$ can be calculated from $\mu_{low}{}^{\dagger}$ and $\mu_{high}{}^{\dagger}$. On the other hand, assuming Z_{tent} is not 13, for example, $Z_{tent} = 7.0$, $Z_{eff,low}$ is calculated as 9.4 and $Z_{eff,high}$ is calculated as 11.8. Although $Z_{eff,low}$ and $Z_{eff,high}$ do not match the proper value, it can be used as a criterion to check that results. The most obvious criterion is to check whether Z_{tent} and the $Z_{eff,low}$ or $Z_{eff,high}$ is a match. Figure 4.9b represents trends for $Z_{eff,low}$ and $Z_{eff,high}$ as a function of Z_{tent}. The dashed line represents the relationship $Z_{tent} = Z_{eff}$; this shows that the analyzed Z_{eff} should be in agreement with Z_{tent} when the corrections for beam hardening and detector response are successful. Furthermore, it is an interesting point that $Z_{eff,low}$ and $Z_{eff,high}$ do not match most Z_{tent}, but they do match when $Z_{tent} = Z_{eff}$. This point is the correct solution. From this figure, we can also understand the following things: (1) Since the results obtained for $Z_{eff,high}$ and that from $Z_{eff,low}$ are in complete agreement, the analytical results for $Z_{eff,\,low}$ are not necessarily required, and (2) the accuracy of the analysis when using $Z_{eff,low}$ is lower than that of $Z_{eff,high}$. This is because the result for $Z_{eff,low}$ is closer to the broken line, and it is difficult to find the intersecting point between the analytical results and the broken

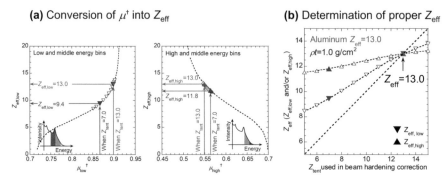

Fig. 4.9 Detail descriptions for (**a**) conversion of normalized μ into Z_{eff}, and (**b**) proper Z_{eff} determination. This figure shows the typical results when aluminum ($Z = 13$) having a mass thickness of 1 g/cm² is selected as an object. We can see that the results for Z_{eff} becomes 13 when Z_{tent} is preliminary determined as 13. By identifying a consistent point, Z_{eff} can be derived

line. It was anticipated that such results would be obtained, because there is only a small amount of information that can be derived for X-ray attenuation from the low energy bin.

Then, we want to analyze the effect of response function on the determined Z_{eff} accuracy. Figure 4.10 is a summary of the plot for identifying Z_{eff} which was estimated by simulation. We prepared the following samples for Z_{eff} simulation: aluminum, acrylic, and a composite material of acrylic and aluminum. The theoretically calculated Z_{eff}s are 6.5 for acrylic, 13.0 for aluminum, 10.5 for composite material in which acrylic and aluminum are mixed in amounts of 0.5:0.5, 9.5 for composite material having composition a ratio of 0.67:0.33, and 8.5 for composite material having composition ratio of 0.8:0.2. Furthermore, in order to consider the effect of thickness, we studied conditions having different mass thicknesses: $\rho t = 1.0$ g/cm², 5.0 g/cm², and 10.0 g/cm². Figure 4.10a shows the results for Z_{eff}s in which the theoretical spectra **IΦ** with absorption of various materials were used. In all the results, we can see that the $Z_{eff,low}$ and $Z_{eff,high}$ intersect Y = X line, and the intersectional point is equal to the theoretical Z_{eff}. This shows that the proposed method can calculate Z_{eff} using both $Z_{eff,low}$ and $Z_{eff,high}$.

On the other hand, as shown in Fig. 4.10b, interesting results were obtained when the X-ray spectra **R⁽¹⁾Φ** was taken into consideration instead of **IΦ**. Regarding the acrylic sample, intersectional points can be clearly confirmed for $Z_{eff,high}$ and $Z_{eff,low}$. However, in the case of aluminum samples and composite materials, it is difficult to

Fig. 4.10 (continued) blue and red closed triangles correspond to acrylic and aluminum, respectively. The green open triangles are bilayer structures of acrylic and aluminum; the ratio of acrylic and aluminum varied as follows, 0.5:0.5, 0.67:0.33, and 0.8:0.2 which are presented in the upper, middle, and lower figures, respectively. The results corresponding to mass thicknesses of 1.0, 5.0, and 10.0 g/cm² are presented in this figure. The determined Z_{eff}s are in good agreement with theoretical values

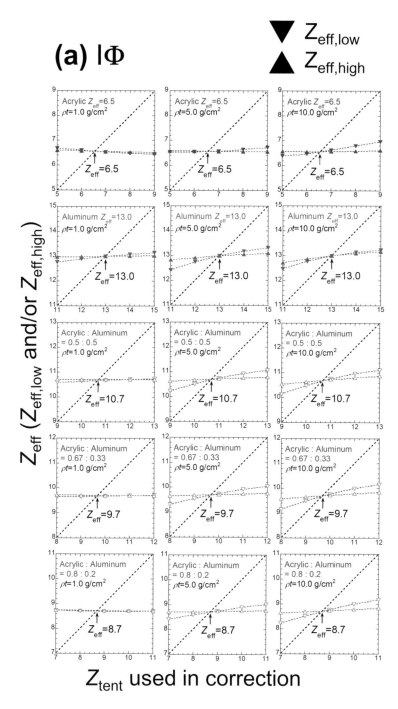

Fig. 4.10 Comparison of Z_{eff} determination curves for (**a**) $\mathbf{I\Phi}$, (**b**) $\mathbf{R^{(1)}\Phi}$, and (**c**) $\mathbf{R^{(1)}R^{(2)}\Phi}$. As a function of Z_{tent}, the $Z_{\text{eff,low}}$, and $Z_{\text{eff,high}}$ are plotted using lower and upper triangles, respectively. The

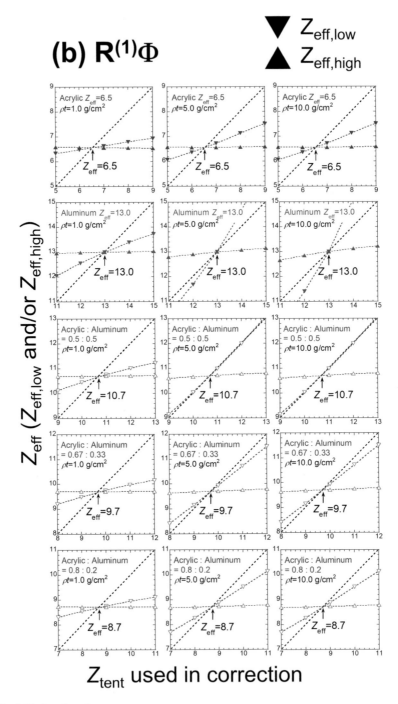

Fig. 4.10 (continued)

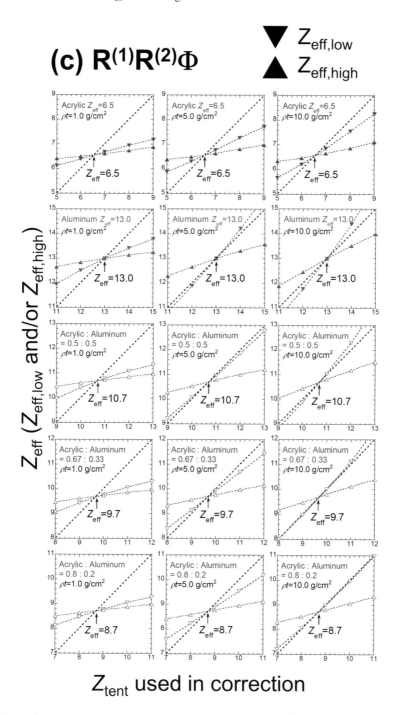

Fig. 4.10 (continued)

identify any intersectional points from intersectional points of $Z_{\text{eff,low}}$, because the trend is similar to Y = X line. This trend becomes more clear when $\mathbf{R}^{(1)}\mathbf{R}^{(2)}\mathbf{\Phi}$ is used as the X-ray spectra as shown in Fig. 4.10c. This fact shows that Z_{eff} can be calculated from the $Z_{\text{eff,high}}$ trend, but the appropriate Z_{eff} cannot be calculated from the $Z_{\text{eff,low}}$ trend. Namely, the information from the low energy bin is disrupted by the effect of the response function, and the attenuation information originally found in the low energy bin is lost. Since our method can correct the charge sharing effect which highly interferes with the low energy bin, when $\mathbf{R}^{(1)}\mathbf{R}^{(2)}\mathbf{\Phi}$ is applied, the intersectional point shows a theoretical value for Z_{eff}. However, in the actual data, the effect of statistical fluctuation is included and each data point deviates. The fact that there is similarity between the trend of $Z_{\text{eff,low}}$ and Y = X line means that this algorithm is difficult to apply to this analysis. From this result, it is a very reasonable method when the energy bins are set at middle (32 keV–40 keV) and high (40 keV–50 keV), because X-ray attenuation information can be derived from these energy regions which are not disrupted by the charge sharing effect and the detector response. In the next section, we only use $Z_{\text{eff,high}}$ ($\mu_{\text{high}}{}^{\dagger}$) for Z_{eff} determination.

Figure 4.11 shows an analytical scheme that provides an effective atomic number image. The left column shows "count images" which are originally measured with an ERPCD. Because our system has three energy bins, the system provides us with three "count images": low, middle, and high energy bins. Here, we also need to measure background images without an object to solve the X-ray attenuation equation. First, a conventional X-ray image can be produced using the following formula:

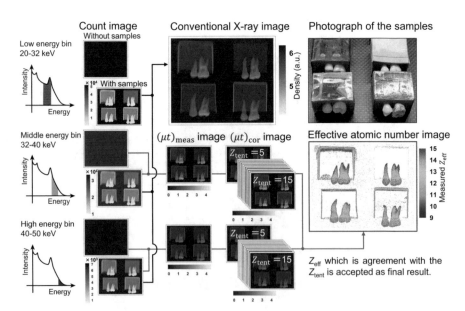

Fig. 4.11 A scheme for creating (1) conventional X-ray image and (2) effective atomic number image from measured X-ray count images using three energy bins

$$\text{Conventional X-ray image} : \log\left(\Sigma N(E) \times E\right)$$
$$= \log\left(N_{\text{low}} E_{\text{eff,low}} + N_{\text{middle}} E_{\text{eff,middle}} + N_{\text{high}} E_{\text{eff,high}}\right), \tag{4.7}$$

where N and E_{eff} are intensity (counts) and effective energy of each energy bin, respectively. An important point is that our system can generate not only a novel image but also a conventional X-ray image. Since the new system can reproduce conventional X-ray images, we expect that novel X-ray images will gradually be accepted clinically as well as the conventional X-ray images created by an ERPCD.

We now turn back to the analysis of the Z_{eff} image. Using the procedure described above, an Z_{eff} image can be reproduced using the count images from the middle and high energy bins. The lower right figure in Fig. 4.11 shows a color map of Z_{eff} dental samples. We can see the fine structure of the tooth. It is clearly seen that the top of the tooth has relatively higher atomic numbers. This trend helps us understand the fact that the top of the tooth consists of enamel which is the hardest tissue in the human body. Although the potential of this analysis is unclear, especially for clinical application, we will continue to do research.

4.3   Experimental Verification of Z_{eff} image

In order to demonstrate how to create an Z_{eff} image, we present some examples. Figure 4.12 represents the experimental arrangement of our test model using an ERPCD. In our system, a line sensor having a pixel size of 200 µm is used. In order to reduce the pulse-pile-up effect and contamination from scattered X-rays, the test model uses a slit scanning system, in which well-collimated X-rays and a line sensor are synchronously moved. Objects are placed on the stage, which is 250 mm away from the detector.

Fig. 4.12 Experimental arrangement of an apparatus to verify our procedure to create X-ray image based on an ERPCD

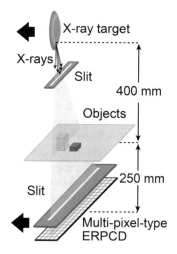

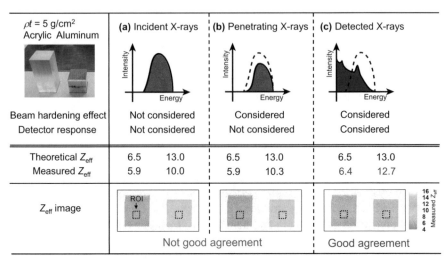

Fig. 4.13 Example of how to analyze an Z_{eff} image under different analytical conditions. It is clearly seen that the beam hardening effect and detector response should be properly corrected

The novelty of our analysis procedure is to correct the beam hardening effect as well as detector response. First, we will explain the importance of applying corrections when μt values are analyzed. Figure 4.13 shows Z_{eff} analysis results for objects which have known effective atomic numbers; in this demonstration, acrylic ($Z_{eff} = 6.5$) and aluminum ($Z_{eff} = 13.0$) having mass thicknesses of 5.0 g/cm² are measured. In the figure, the upper row shows schematic drawings of X-ray spectra which have been analyzed. The lower row shows the Z_{eff} image results. Figure 4.13a shows the results in which corrections for the beam hardening effect and detector response are not carried out. The theoretical X-ray spectra without considering these effects are used to obtain a reference value for μ_{high}†. The result for Z_{eff}s is not in good agreement with the theoretical values. Figure 4.13b shows the results in which correction of the beam hardening effect is only considered when the correction for $(\mu t)_{meas}$ to $(\mu t)_{cor}$ is performed. Although the beam hardening effect in the object is properly corrected, the Z_{eff} results are not in agreement with the theoretical values. On the other hand, Fig. 4.13c shows the results proposed in this book; the effect of detector response in addition to the beam hardening effect is taken into account. The results are in good agreement with the theoretical values. It is clearly seen from these results that the effect of detector response is much larger than that of beam hardening effect. That is why we think it is important to consider the response functions of an imaging detector.

Figure 4.14 shows the effective atomic number image created by measuring single and composite substances composed of acrylic and aluminum. As shown in the photograph at the top of the figure, the mass thicknesses of a series of samples are (a) 1.0 g/cm², (b) 5.0 g/cm², and (c) 10.0 g/cm². We prepared multiple samples with different ratios of acrylic and aluminum to demonstrate the experiment. The second-

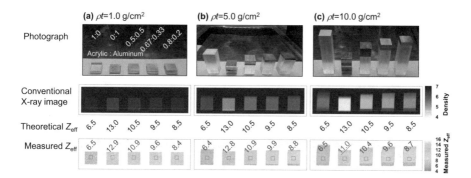

Fig. 4.14 X-ray images produced by our slit-scanning-system. (**a**), (**b**), and (**c**) correspond to objects having mass thicknesses of 1.0, 5.0, and 10.0 g/cm², respectively. In each picture, acrylic, aluminum, and composite substances were arranged in order from left to right. Z_{eff}s of these objects are well known; the theoretical values of these objects are Z_{eff}s of 6.5, 13.0, 10.5, 9.5, and 8.5 regardless of mass thickness. The middle figures are conventional X-ray images. The bottom figures are Z_{eff} images. The ROI and the mean value for Z_{eff} are presented in the figure. For most objects, mean values are within the range of ± 0.5 from the theoretical Z_{eff}, and these values are shown in blue. The value which is out of the ± 0.5 range from the theoretical Z_{eff} is shown in red

row image is a conventional X-ray image for each sample, and the values of total amount of energy absorbed in the pixel are expressed as the image density. Namely, it corresponds to the image when ERPCD is operated as an EID. Roughly speaking, samples with the same mass thickness are expressed in a similar shade. However, since X-ray attenuations related to different atomic numbers are greatly influenced, a small amount of X-ray absorption will occur when aluminum, which has a high atomic number, is used; that is why the aluminum sample is shown with an image density close to white. Because the image density of a conventional X-ray image is determined by two parameters, the mass thickness and the atomic number of the object, it is very difficult to estimate the atomic number of a substance from these images. The image shown in the bottom row is the result of calculating Z_{eff} using ERPCD. We can see that the obtained values are in good agreement with the theoretical estimation. These values can be determined by considering the effects of beam hardening and response functions and using atomic number identification algorithms. The experimental results imply that these physical considerations were correct. On the other hand, there is the case where our algorithm does not work properly. In the example shown in Fig. 4.14, the aluminum sample with $\rho t = 10.0$ g/cm² does not show a true value. A large amount of X-rays are absorbed by the aluminum object and only a small amount of X-rays are incident to the detector. On the other hand, many X-rays without absorption are incident to the air layer which was just placed next to the aluminum object. In our method, the response function is simulated by assuming the situation when the pixel of interest and surrounding pixels are exposed to similar X-ray irradiations, but this condition does not match with the experiment. Therefore, our algorithm did not work properly. We consider that

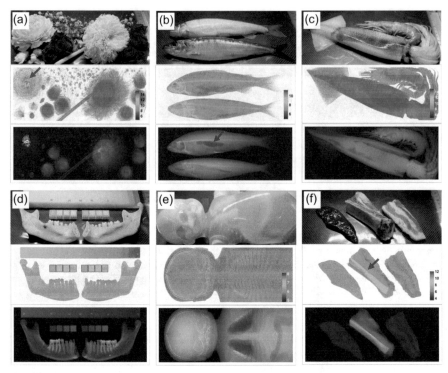

Fig. 4.15 Typical example of images. Photograph, Z_{eff} image and conventional X-ray image are presented

such cases are rare in actual medical applications, but we are currently conducting further studies in order to construct algorithms that can be applied to general cases.

We finally present Z_{eff} images of several samples. We are currently analyzing the results of the Z_{eff} images. We would like to show the results of various samples and conclude them in this book.

Figure 4.15a shows the results of X-ray images of fresh and artificial flowers. From the optical camera image in the upper figure, we can see the fresh flowers and artificial flowers arranged in a similar manner, and it is difficult to distinguish which is the artificial flower using this image. The lower figure shows a conventional X-ray image measured with our ERPCD. We can perceive something like a wick with white on the left side. It is difficult to determine what it is. On the other hand, as shown in the middle figure, the Z_{eff} image can identify foreign materials. In this case, fresh flowers consist of C, H, O, and N atoms; therefore, Z_{eff} of the fresh flowers is between 6 and 7. In contrast, the artificial flower has metallic parts especially in the center core and we can easily identify that the core parts have a higher Z_{eff} of between 12 and 14. Aluminum ($Z = 13$) is one of the most useful parts used.

Figure 4.15b shows the results for fish. In each figure, the upper and lower fish are freshwater and saltwater fish, respectively. From the optical camera image shown on the top it can be clearly understood what it is and does not need to be explained. The lower figure shows a conventional X-ray image, and we can see freshwater fish has a large floater, which is represented in black; in the X-ray path including the gas bladder has a smaller X-ray attenuation and it results in more X-ray penetration. When seeing the middle Z_{eff} image based on understanding the conventional X-ray image, the Z_{eff} image shows an interesting trend. In the Z_{eff} image, a difference in bone structure is clearly observed, but notably we cannot identify the floater. Since the air region contributes little to the attenuation of X-rays, the influence of the air region does not contribute significantly to the calculation of Z_{eff} using the attenuation calculation for X-rays. This is very important for understanding an Z_{eff} image. In other words, it is important to understand that the Z_{eff} image is not a coloration of a conventional X-ray image, and that the color reflects the physical phenomenon of X-ray attenuation.

Figure 4.15c shows the results for shrimp and squid. We think these kind of objects are difficult to analyze when applying Z_{eff} analysis. Although some fine structures can be derived from the Z_{eff} image of shrimp and from the Z_{eff} image of squid, it is difficult to read detailed information.

Figure 4.15d shows the results of the jawbone and teeth. In this demonstration, actual extracted bone and a phantom to measure bone mineral density (BMD) are simultaneously measured. A BMD phantom was used to find the same image density of a known reference within the analysis region. In the lower figure, the color range of the BMD phantom and the conventional X-ray image seems to be in agreement; namely, this phantom can be applied to measure BMD for actual diagnosis using conventional X-ray examination. However, as shown in the middle figure, the range of Z_{eff} is completely different for bone (approximately $Z_{eff} = 13$) and BMD phantom ($Z_{eff} = 6–12$).

Figure 4.15e shows the results for the head and chest regions of a neonate phantom. Large differences between the conventional X-ray image and Z_{eff} image can be observed in the lung region. The Z_{eff} image cannot differentiate the lung regions; as explained above, the Z_{eff} image cannot discriminate void regions filled with air. On the other hand, precise bone structure of the head and chest regions can be clearly identified in the Z_{eff} image. As shown here, even if a medical image is obtained, it is not easy to evaluate the diagnostic advantages of using an Z_{eff} image. We hope that the utilization of Z_{eff} images will be found in X-ray diagnostics.

Figure 4.15f shows a comparison of meat; left, middle, and right are liver, spare rib, and pork ribs. As shown in the conventional X-ray image, the spare rib has high X-ray attenuations, but it is unclear what causes the X-ray attenuations. By comparing this image, the Z_{eff} image expresses a different trend. From the Z_{eff} image, we can identify the bone with high Z_{eff} materials shown in yellow to red, and muscle and fat are represented by green and blue, respectively.

References

1. T.E. Jhonson, B.K. Birky, *Health Physics and Radiological Health* (Lippincott Williams & Wilkins, Philadelphia., ISBN-10: 1609134192, 2011)
2. B.J. Heismann, J. Leppert, K. Stierstorfer, Density and atomic number measurements with spectral X-ray attenuation method. J. Appl. Phys. **94**, 2073–2079 (2003). https://doi.org/10.1063/1.1586963
3. F. Tatsugami, T. Higaki, M. Kiguchi, S. Tsushima, A. Taniguchi, Y. Kaichi, T. Yamagami, K. Awai, Measurement of electron density and effective atomic number by dual-energy scan using a 320-detector computed tomography scanner with raw data-based analysis: a Phantom study. J. Comput. Assist. Tomogr. **38**(6), 824–827 (2014). https://doi.org/10.1097/RCT.0000000000000129
4. J. Fornaro, S. Leschka, D. Hibbeln, A. Butler, N. Anderson, G. Pache, H. Scheffel, S. Wildermuth, H. Alkadhi, P. Stolzmann, Dual- and multi-energy CT: approach to functional imaging. Insights Imaging **2**, 149–159 (2011). https://doi.org/10.1007/s13244-010-0057-0
5. A.C. Silva, B.G. Morse, A.K. Hara, R.G. Paden, N. Hongo, W. Pavlicek, Dual-energy (Spectral) CT: applications in abdominal imaging. Radiographics **31**(4), 1031–1046 (2011). https://doi.org/10.1148/rg.314105159
6. S. Faby, S. Kuchenbecker, S. Sawall, D. Simons, H.-P. Schlemmer, M. Lell, M. Kachelrieß, Performance of today's dual energy CT and future multi energy CT in virtual non-contrast imaging and iodine quantification: a simulation study. Med. Phys. **42**(7), 4349–4366 (2015). https://doi.org/10.1118/1.4922654

Chapter 5
Summary

In this book, we proposed how to create an effective atomic number (Z_{eff}) image using an Energy-Resolving Photon Counting Detector (ERPCD). The theory to identify an object is based on the analysis of X-ray attenuations. We showed the necessity of taking into consideration the beam hardening effect and detector response when analyzing X-ray attenuations measured with an ERPCD. We carefully described the calculation method used for the response function, and the magnitude of the effect on the analysis results. In particular, in a study using simulations, we revealed that the X-ray attenuation information originally contained within an energy bin of interest may be lost due to the influence of response function on the spectrum. Then, we proposed a correction method for extracting X-ray attenuation information appropriately. However, this method requires tentatively determining the atomic number of an object. Therefore, we devised a methodology that can simultaneously determine the atomic number Z assumed to be used for correction and finally determined an effective atomic number Z_{eff}. In order to demonstrate our novel procedure, experiments using a test model of an ERPCD were performed, and some typical Z_{eff} images are presented as well as conventional images.

Figure 5.1 shows an educational poster describing a photon counting detector presented at the International Conference (Radiological Society of North America: RSNA2019). Fortunately, we were able to win the "certificate of merit award" in this poster presentation, but it was deemed too long to be published as an original research paper. Therefore, the information used to create this poster was re-edited and put together in this book. We hoped that the contents gain sympathy as the educational poster which will be re-readable in this book, and will provide useful knowledge when searching for the application of a photon counting detector.

Although we presented precise analytical procedures for creating Z_{eff} images, it is unclear whether or not Z_{eff} images can be directly applied to medical diagnosis. Further studies should be carried out to obtain good clinical results. Photon counting techniques have flaws; we should reduce the pile-up effect and contamination from scattered X-rays. The examples given in this book only demonstrate the

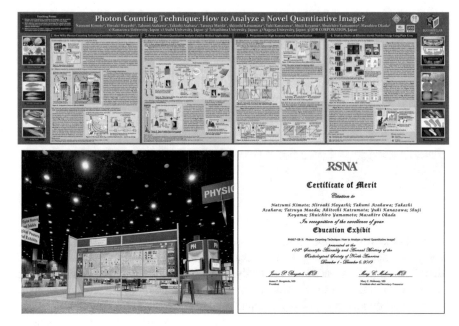

Fig. 5.1 Poster presentation at RSNA2019. The contents in this poster are partially used in this book

methodology under ideal conditions. In parallel with technological innovation, additional research should be performed before photon counting techniques can be used in clinical situations.

Acknowledgements The descriptions in this chapter were partially supported by a collaborative research between Kanazawa University and Job corporation (https://www.job-image.com/), Japan. The author wishes to express gratitude to Dr. Shuichiro Yamamoto, Mr. Masahiro Okada, Mr. Fumio Tsuchiya, Mr. Daisuke Hashimoto, Mr. Yasuhiro Kuramoto, and Mr. Masashi Yamasaki for their valuable contributions. Also, we would like to thank to Mr. Tsutomu Yamakawa who belongs to Rad Imaging regarding the establishment of a research team due to his centripetal efforts and valuable discussions in the early stages of our research. The clinical application of a photon counting detector was also discussed with Dr. Yoshie Kodera and Dr. Shuji Koyama who belong to Nagoya University, Japan.

We are very grateful to Dr. Krzysztof Iniewski who is managing R&D at Radlen Technologies, Inc. for giving us this great opportunity to write this book. We collaborated with him on a previous book [1], and we wrote this book as a modified version of the previous book, with much larger content.

Reference

1. S. Reza, K. Iniewski, et al. (eds.), *Semiconductor Radiation Detectors* (CRC Press, Boca Raton., 290 pages, ISBN: 9781138710344, 2017)

Index

A
Artificial intelligence (AI), 35
Atomic model, 4–6, 8
Atomic number, 94
Auger electron emission, 54, 55
Avogadro number, 94

B
Beam hardening corrections, 88–90, 101–103
Beam hardening effect, 86, 87, 89, 90, 100,
       101, 110
Birch's formula, 64
Bohr's model, 6
Bone mineral density (BMD), 41, 113
Bremsstrahlung X-ray, 4, 16, 18, 19, 25

C
Cadmium telluride (CdTe) detector, 61
   polychromatic X-ray spectra, 63
   Response function, 45, 59
   single-probe-type CdTe detector, 61
   X-ray spectroscopy, 62
Cadmium zinc telluride (CZT) detectors, 50
   multi-pixel-type CZT detector, 52
   response function, 64, 68
   and Si, 51
   X-ray spectra, 72–81
Charge cloud, 48
Charge sharing effect, 49, 70, 72, 74, 77, 81
Charge transport process, 48, 72
Compton escapes (CE), 59, 60, 62, 66
Compton scattering effect, 55, 94
Computed tomography (CT) imaging, 29

Conventional unfolding method, 48
Conventional X-ray image, 32–34
Count image, 108, 109

D
Detector response function, 45, 46
Diagnosis using X-rays, 27–29
Dual-energy CT, 40, 99
Dual-energy X-ray absorptiometry (DEXA), 41
Dual X-ray technique, 36

E
Effect of detector response, 101, 102, 110
Effective atomic number (Z_{eff}), 34, 39
     derivation using ERPCD, 108
       images, 108–113
       method, 100–109
Energetic electrons, 13–16
Energy band, 8
Energy integrating detector (EID), 36,
       37, 99, 111
Energy resolution, 72–73
Energy state, 6
Energy resolving photon counting detector
       (ERPCD), 36–39, 115
   attenuation factor measurement, 81–83
   educational poster, 115, 116
   medical examinations, 50
   response function, multi-pixel-type ERPCD
       (*see* Multi-pixel-type CZT detector)
   Z_{eff} value (*see* Effective atomic number
       (Z_{eff}) image)
Escape peaks (EP), 60–63, 66–68, 71, 76

F
Frequency condition, 5
Full energy peak (FEP), 59–63, 66, 79

G
General radiography system, 93

H
Half-value layer (HVL), 84
Hydrogen atom, 5

I
Image analysis method, 36
Image reconstruction, 93
Ionization energy, 9–10
Irradiatin field, 64–66

K
Klein-Nishina formula, 57
Kramers' formula, 19

L
Leaf electroscope, 1–2
Linear attenuation coefficient
        (μ), 38, 99, 101
    normalized linear attenuation
        coefficient, 101

M
Mass attenuation coefficient, 95
Medical applications
    analysis of objects, 40–42
    clinical diagnosis, 34
    diagnosis, 34–36
    EID, 36, 37
    equipment, 29, 31
    ERPCD, 36–39
    gray-scale image, 33
    material identification, 31, 42, 43
    penetrating X-rays, 34
    pixel size, 32
    principles, 31
    two-dimensional digital sensor, 32
Medical X-ray, 24
Monochromatic X-ray, 101
Monte-Carlo simulation, 49, 60, 66, 68, 69
Movable diaphragm, 23–24
Multi-channel-analyzer (MCA), 48

Multi-pixel-type CZT detector, 52
Multielectron atom, 8–10

N
Normalized linear attenuation coefficient,
        101, 102

P
Pediatric table, 11
Photoelectric effect, 53, 94
    Cd and Te atoms, 50
    energy E, X-ray, 54
    and succeeding phenomena, 53
    X-ray emissions, 50
Photon counting detector (PCD), 35, 96, 99
    educational poster, 115, 116
Photon counting technique, 45, 115
Photons, 11–13
Pixel size, 68–70
Polychromatic X-rays, 25, 45, 59, 63, 69–72,
        80–81, 101, 102
Procedure to make medical X-rays, 36

Q
Quantum condition, 5
Quantum mechanics, 3

R
Radiation, 1
Radiation measurement, 9
Radiation physics, 6
Response function, 45
    charge collecting process, 48, 49
    charge sharing effect, 72
    charge transporting process, 72
    concept, 46
    energy resolution, 72, 73
    full energy absorption event, 46
    multi-pixel-type ERPCD, 48, 64–71
    single-probe-type CdTe detector, 59–64
    unfolding correction, 46–48

S
Slit scanning system, 109, 111
Spectrum, 23–26

U
Unfolding correction, 46–48, 62–64

X
X-rays
  application, 27–29
  attenuation, 21–22
  Cd and Te characteristic X-rays, 55–56
  characteristic X-rays, 17, 18
  electro magnetic waves, 3, 4
  equipment, 3
  hydrogen atom, 3
  industrial and clinical applications, 1
  ionization, 1
  mechanism of generation, 3
  polychromatic X-rays, 18–20
  reproduction of measured X-ray
    spectra, 74–79
  spectrum, 23–26
  target, 25–26
  target material, 13–16
  visible light, 3

Printed in the United States
by Baker & Taylor Publisher Services